**Organic Principles and Practices Handbook Series**
A Project of the Northeast Organic Farming Association

# Organic Dairy Production

**Revised and Updated**

SARAH FLACK

Illustrated by Jocelyn Langer

CHELSEA GREEN PUBLISHING
WHITE RIVER JUNCTION, VERMONT

Editorial Coordinator: Makenna Goodman
Project Manager: Bill Bokermann
Copy Editor: Cannon Labrie
Proofreader: Helen Walden
Indexer: Peggy Holloway
Designer: Peter Holm, Sterling Hill Productions

Printed in the United States of America
First Chelsea Green revised printing March, 2011
10 9 8 7 6 5 4 3 2 1   11 12 13 14

Chelsea Green Publishing is committed to preserving ancient forests and natural resources. We elected to print this title on 30-percent postconsumer recycled paper, processed chlorine-free. As a result, for this printing, we have saved:

**4 Trees (40' tall and 6-8" diameter)**
**1 Million BTUs of Total Energy**
**378 Pounds of Greenhouse Gases**
**1,822 Gallons of Wastewater**
**111 Pounds of Solid Waste**

Chelsea Green Publishing made this paper choice because we and our printer, Thomson-Shore, Inc., are members of the Green Press Initiative, a nonprofit program dedicated to supporting authors, publishers, and suppliers in their efforts to reduce their use of fiber obtained from endangered forests. For more information, visit: www.greenpressinitiative.org.

Environmental impact estimates were made using the Environmental Defense Paper Calculator. For more information visit: www.papercalculator.org.

**Our Commitment to Green Publishing**

Chelsea Green sees publishing as a tool for cultural change and ecological stewardship. We strive to align our book manufacturing practices with our editorial mission and to reduce the impact of our business enterprise in the environment. We print our books and catalogs on chlorine-free recycled paper, using vegetable-based inks whenever possible. This book may cost slightly more because we use recycled paper, and we hope you'll agree that it's worth it. Chelsea Green is a member of the Green Press Initiative (www.greenpressinitiative.org), a nonprofit coalition of publishers, manufacturers, and authors working to protect the world's endangered forests and conserve natural resources. *Organic Dairy Production* was printed on Joy White, a 30-percent postconsumer recycled paper supplied by Thomson-Shore.

**Library of Congress Cataloging-in-Publication Data**
Flack, Sarah, 1969-
Organic dairy production / Sarah Flack ; illustrated by Jocelyn Langer. -- Updated and rev.
p. cm. -- (Organic principles and practices handbook series)
Originally published in 2004 by the Northeast Organic Farming Association Interstate Council.
Includes index.
ISBN 978-1-60358-351-0
1. Organic dairy farming--United States. 2. Organic dairy farming--Case studies. I. Langer, Jocelyn. II. Title. III. Series: Organic principles and practices handbook series.

SF246.O74F63 2011
636.2'142--dc22

2011003649

Chelsea Green Publishing Company
Post Office Box 428
White River Junction, VT 05001
(802) 295-6300
www.chelseagreen.com

## Best Practices for Farmers and Gardeners

The NOFA handbook series is designed to give a comprehensive view of key farming practices from the organic perspective. The content is geared to serious farmers, gardeners, and homesteaders and those looking to make the transition to organic practices.

Many readers may have arrived at their own best methods to suit their situations of place and pocketbook. These handbooks may help practitioners review and reconsider their concepts and practices in light of holistic biological realities, classic works, and recent research.

Organic agriculture has deep roots and a complex paradigm that stands in bold contrast to the industrialized conventional agriculture that is dominant today. It's critical that organic farming get a fair hearing in the public arena—and that farmers have access not only to the real dirt on organic methods and practices but also to the concepts behind them.

## About This Series

The Northeast Organic Farming Association (NOFA) is one of the oldest organic agriculture organizations in the country, dedicated to organic food production and a safer, healthier environment. NOFA has independent chapters in Connecticut, Massachusetts, New Hampshire, New Jersey, New York, Rhode Island, and Vermont.

This handbook series began with a gift to NOFA/Mass and continues under the NOFA Interstate Council with support from NOFA/Mass and a generous grant from Sustainable Agriculture Research and Education (SARE). The project has utilized the expertise of NOFA members and other organic farmers and educators in the Northeast as writers and reviewers. Help also came from the Pennsylvania Association for Sustainable Agriculture and from the Maine Organic Farmers and Gardeners Association.

Jocelyn Langer illustrated the series, and Jonathan von Ranson edited it and coordinated the project. The Manuals Project Committee included Bill Duesing, Steve Gilman, Elizabeth Henderson, Julie Rawson, and Jonathan von Ranson. The committee thanks SARE and the wonderful farmers and educators whose willing commitment it represents.

ACKNOWLEDGMENTS

Thanks to Carl Reidel, Amy McMillon, Remi Gratton, Carol Dunsmore, Gwyneth Flack, Steve St. Onge, and Doug Flack for the editorial help. Special thanks also to Nat Bacon and Lisa McCrory, and to the many farmers and veterinarians who provided me with information and inspiration.

# CONTENTS

# Introduction

## The Decision to Be Organic

This handbook is an introduction to important issues in organic dairy production. It attempts to provide today's organic or transitioning dairy farmer with an overview of the tools and ideas available. It includes resources to assist farm management and ongoing farm improvement. Real-farm examples are included to show the variation and creativity in how different organic dairy farms are managed.

There is no simple recipe for setting up and managing a successful organic dairy farm; the uniqueness of people alone assures that there are as many ways to go as there are organic dairy farmers. The success of each depends on the way the farmer manages the available natural and human resources to meet the particular farm, family, and business goal.

Even knowing products, information, and resources is just a beginning. The ability to manage effectively requires technical knowledge, gained over time, of managing and caring for soils, livestock, and plants. It requires observational skills to notice subtle changes and adapt to them. The decision to go organic, a transition that takes several years, needs to be considered within the context of the farmer's experience and predisposition, the overall goals for the farm, the family goals, the natural ecology of the land involved, and an understanding of the organic standards.

Before making the transition to organic dairy farming, ask yourself:

- Do you have management skills in livestock health, soil fertility, high-quality forages, record keeping that meets organic standards, and maintaining high milk quality?
- Do you have access to an organic milk market?
- Do you have a clear understanding of the current organic standards?
- Are there locally available and affordable sources of organic feed? Do you know where to source alternative health-care products and organically approved fertilizers?

- Do you know the costs associated with transition and long-term organic production? The cost of the transition is a key consideration for a farm deciding if and when to convert to organic.
- Do you know people in your area who are knowledgeable (veterinarians, agronomists, other farmers)?
- Do you have a clear (and written) goal for your family and farm? (See the companion NOFA handbook *Whole-Farm Planning* by Elizabeth Henderson and Karl North.)

Several decades ago, the business of organic dairy farming had its beginnings with just a few farms. These farms were organic owing to personal commitment or because they had a locally developed market. Today the picture is quite different. Growing numbers of certified organic dairy farms are shipping fluid milk or making organic dairy products on their farms.

Many of the early farmers who transitioned to organic dairy production had already adopted management-intensive grazing. For them the switch was a matter of changing some soil-fertility inputs and health-care treatments and setting up a record-keeping system to meet organic standards. Now there are increasing numbers and types of dairy farms making the transition to organic. This rapid growth is due to the higher prices being paid for organic milk and widening knowledge of organic methods of livestock and crop management. It is also due to many farmers' dissatisfaction with the practices in high-production conventional farming, with its low financial returns and swings in milk prices.

Dairy farmers making the transition to organic production today have some advantages over farmers who did so ten years ago. They enjoy a wider array of approved health-care products, more sources of organic seeds and soil amendments. They have access to more veterinarians and soil and crop specialists with knowledge of organic practices and more organic-farmer peers to learn from. There is also more access to markets for organic fluid milk.

Challenges, however, remain for organic dairy farmers and those making the transition. In some locations it is difficult to find an organic milk market. It may be challenging to find a knowledgeable veterinarian or crop or soil advisor. Additionally, the increasing number of organic

> Organic dairy farming is not just conventional dairy farming without the antibiotics, synthetic fertilizers, and pesticides. Establishing a sustainable organic dairy farm is far more than substituting "organic" products for synthetic ones. Rather, it is an ongoing process of creating a complex and diverse agricultural ecosystem. It involves continued work to improve the health of soils, quality of forages, vitality of livestock, and the quality of life of the farm family. It is a new way of seeing and understanding agriculture in all its aspects.

livestock health-care products requires learning how to use them, which ones are allowed under organic standards, and which are effective.

## Cheap Food and Organic Milk

> Organic dairy farming is big business. It is no longer a radical fringe that attracts only the most innovative and risk taking farmers.
> —Hubert Karreman, DVM, *Treating Dairy Cows Naturally*

What will happen to the price paid to farmers for organic milk in the long run is a matter of ongoing discussion. As a growing number of organic dairy farms add to the supply, will demand continue to grow? Will milk buyers and processors continue to pay organic dairy farmers fairly? Will organic milk become another cheap commodity food and drive down the price to farmers?

Opinions vary, but all agree that the situation should be watched closely and that organic dairy farmers will benefit from sticking together and staying informed. The Northeast Organic Dairy Producers Alliance (NODPA) can help dairy farmers stay in communication with each other and stay informed about the complex, changing world of milk marketing and the wide array of information about organic dairy production. The NODPA Web site (www.nodpa.com) lists other national and regional organizations that help organic dairy farmers stay informed and connected.

The long-range solution to creating an organic dairy system that will sustain family farms will come from recreating local food systems—market arrangements where milk isn't shipped long distances to the processing facility and from there long distances to the consumer. The job of creating this food system is already under way, and will require much work by consumers, advocacy groups, and farmers.

one

# Organic Certification: Why, How, and What?

If you decide to begin the process of getting your farm certified, it is essential to gain a clear understanding of both the process and the standards. The information on the organic standards and certification process needs to come from the National Organic Program (NOP) or the certifier. Don't rely on your grain dealer or veterinarian, as much secondhand information about the organic standards and certification process is inaccurate.

Before you apply for certification, here is a list of things to consider:

- Determine if the market you are considering actually requires certification.
- Secure a market. Certification does not guarantee a market. Become familiar with the marketing options, market stability, and market prices in your area.
- Become familiar with the organic standards and transition process.
- Be sure your record-keeping system is adequate.

Once you decide you want to pursue certification, you'll need to choose a certifier. There is a list of organic certification organizations on the National Organic Program Web site (http://www.ams.usda.gov/AMSv1.0/nop). On the NOP Web site you can also find a copy of the full organic standards and other information. Once you choose a certifier, you'll want to contact them to begin the process of certifying the land and livestock. They'll send you their guidelines and fee information, as well as a copy of the application forms. The certifier may require a visit by someone to determine if you are ready to begin the transition process, which may involve an additional fee. Land requires a three-year transition, livestock one year. You will need to keep good records. Finding a mentor for this process can be helpful.

It is important to speak with the certifier before you begin the transition process; some certifiers have an initial application/inspection process for livestock. Once your final transition year is mostly complete and you have sent in your completed application and any additional information they request, the certifier or inspector will contact you to set the date of your annual inspection. The timing, steps, and cost vary among certifiers.

During the first inspection, expect your inspector to visit all the fields and pastures, buildings, and animals. The inspector will look at buffers to adjoining land and assess potential contamination risk from neighboring farms and the risks of what you produce being commingled with nearby nonorganic products. The same assessment applies to the livestock facilities, animals, feed storage, fertilizers, and pest-management products stored or used.

The inspection will cover all the livestock health-care products, milking system, product storage, all cleaning products, and any processing or sales areas. The review of records will include health-care records, crop records of harvest and spreading of manure and other amendments, sales and purchase receipts for livestock, health-care products, fertilizers, pest products, milk, and feed. The inspector will look at your pastures and your feeding records to verify that you are providing enough dry matter (forage after the moisture has been removed) from pasture to meet the organic pasture requirement. They will also verify that your housing and outdoor access areas allows all animals over six months of age daily outdoor access during the non-grazing season.

The inspector's job is to verify that the farm is operating according to the organic management plan as described in your application. It is not the inspector's job to make the certification decision. The inspector is prohibited by the regulations from giving the farmer any advice that may help him or her overcome any noncompliance issues or potential barriers to certification.

At the end of your farm inspection,

the inspector should review the visit with you in an exit interview. This should include the list of any potential noncompliance issues that came up during the inspection.

After your inspection visit, the inspector returns your application with a report to the certifier. The certifier reviews the material and makes a determination on your farm. If they decide that your farm qualifies, you will be notified and receive a copy of the certification for your farm. The certifier may grant the certification contingent on some management changes. The certifier sometimes denies certification owing to some major management changes that need to be made. If you don't agree with the decision, the appeals process is described in the National Organic Program (NOP) standards. Additional information on the transition and certification process is available from several organizations including NODPA, eOrganic, ATTRA, and other resources listed at the end of this book.

All organic farmers need to have a copy of the organic standards. You can get these online, or you can get a copy from your certifier. In general, the standards specify the following:

- No synthetic fertilizers or pesticides can be used on any pasture or crop land.
- Ruminants (cows, sheep, goats) over six months old must be pastured during the grazing season. During that time they must get an average of 30 percent of their dry-matter intake from pasture.
- Ruminants over six months old must have daily outdoor access during the non-grazing season.
- Calves must be fed organic milk.
- Tail docking is prohibited.
- All livestock must get 100 percent organic feed, including grains, forages, pastures; any supplemental minerals or vitamins must contain only approved ingredients.
- Only approved health-care products may be used. Most conventional health-care products such as antibiotics, hormones, steroids, etc., are prohibited.
- Only approved fly-control methods and products may be used.

- Keep required records (crops, sales, purchases, feeds, livestock, and so on).

Keep in mind that this is not a comprehensive list!

### FEATURED FARM:
### Spring Brook Farm, Westfield, Vermont, Lyle (Spud) and Kitty Edwards

(Excerpt reprinted with permission from an article by Nat Bacon of NOFA–Vermont in *NODPA Notes* 3, no. 3.)

Before transitioning, Spud was worried about managing mastitis without antibiotics, but quickly came to feel that alternative treatments were about as effective as conventional mastitis tubes, and that mastitis and animal health were better under a less stressful system. Tired of the direction of conventional dairying, "I was going to sink or swim with organic," Spud recalls. He feels that the popular notion that cows won't milk well under organic management is a myth, and that maintaining strong genetics and putting up good forages are the keys to making milk profitably.

Spud says that he farms according to the KISS philosophy—Keep it simple. He milks 50 Holsteins in a tie-stall barn, although he would eventually like to cut down to 40 cows and hire less labor. The barn can hold all the milkers and young stock, although Spud would like to build a greenhouse-type barn to improve animal comfort and reduce crowding in the main barn. He believes that taking excellent care of his cows results in good milk production, and has the numbers to back it up: 11 of 43 cows milked over 100 pounds on a recent DHIA test day, and the herd maintains an 18,500-pound rolling herd average. Spud tries to keep as much good grass as possible in front of the cows, although when the pasture gets short he supplements around one-third of the ration with high-quality baleage. In the winter, he cuts the bales with a tractor-mounted hay slicer in the haymow, and forks it down to the cows below. The barn has two grain bins; one holds straight cornmeal and the other a higher-protein mix. Instead of changing the grain constantly, Spud just adjusts the amount of protein mix he feeds his cows, depending on the pasture quality and time of year. Grain amounts are figured at 1:3 or 1:4 ratio of grain to milk, depending on the stage of lactation, and are fortified with kelp and minerals. Buying just two grains allows Spud to feed his young stock milker grain, without needing an extra bin.

In keeping with his philosophy, Spud has a very manageable cropping program on 45 pasture acres and 35 hay acres. He gives cows a fresh paddock daily, and with timely hay supplementation can keep lush grass through the season. Spud thinks Orleans County, with its usually consistent rain showers, is a good place to graze cows. He gets a good response from adding Sul-Po-Mag to his fields, and liming when needed.

In terms of herd health, Spud and Kitty (who does a lot of the herd health work) are big believers in preventative measures and boosting the cows' immune systems. They keep vaccinations up to date, including Lepto and rabies shots. For animals that look off-health, Spud likes using the Crystal Creek drenches and the Impro capsules sold through Brookfield Ag Services. If a cow comes down with mastitis, Spud treats with aspirin, vitamin C shots, and peppermint ointment, and he keeps the infected quarter stripped out. He feels it's important to use a good barrier teat dip, to give selenium/vitamin E boosters, and to dry cows down to 30 pounds of milk per day at dry-off to prevent problems. (Note that some barrier teat dips may not be allowed by your local certifier—check before using!) Attention to detail results in good udder health: a 71,000 somatic-cell count average on the most recent DHIA test. Under organic management, Spud feels cows are just less stressed and healthier. His cull rate is 12 to 15 percent, about half of the conventional cull rate. Keeping animals healthy while milking strongly is the key to Spring Brook Farm's success.

---

Healthy soil = healthy cow = healthy farm = healthy people. Soil health not only helps livestock thrive, it provides high-quality forages that cows convert into nutrient-dense meat and milk, which support human health. Such products, when sold to a discerning customer, can also support the farm financially.

two

# Soil—The Foundation of Health

Improving the care given to the soil can prevent many cow health problems. Dairy farms with soils that have correctly balanced fertility, high organic matter, and good biological activity will produce higher-quality stored forages and improved pastures. Farms with healthy soils growing more nutritious, higher-quality forages should show higher milk yields while feeding fewer concentrates. Some farmers report management of internal parasites, which can be a challenge with young stock on organic dairy farms, is easier after improving soil health. They also report experiencing fewer metabolic and other cow health problems, having longer-lived animals and more replacements reaching productive years.

Part of the job of organic dairy farming is creating a farm in which microorganisms in both the soil and in the livestock's digestive system are abundant and healthy. These organisms play an essential role in cycling nutrients and energy from soil, air, and sun to plant, animal, and manure and back to the soil.

An organic dairy farm, unlike other types of farms, does not grow only crops or keep only livestock. It does both. The availability of manure produced on-farm, and the fact that much of the farm is in sod (in some cases the entire farm), creates both challenges and opportunities in the development of a sustainable approach to soil fertility.

Healthy soils produce the quality forages that, as the basis of the dairy cow's diet, allow livestock to reach their potential as healthy parts of the whole farm system. In return, the cattle will produce manure, which if managed well, will be central to the healthy cycling of nutrients on the farm. Soil health not only supports healthy livestock, it provides high-quality forages that cows convert into nutrient-dense meat and milk that support human health. These are products that, when sold to a discerning customer, can also support the farm financially. (See chapter 5 on marketing and chapter 8 on the nutritional qualities of grass-fed milk and dairy products.)

There is no single recipe for creating a soil that is healthy biologically, physically, and chemically. Each farmer must creatively use the practices available within the generally understood guidelines of good management and apply an understanding of how to promote soil vitality. Find a good soil consultant or a mentor, choose some appropriate methods of monitoring soil health, study the basic principles of soil fertility, and refer to books and other resources in the soils section of the resource list at the end of this handbook.

The essential elements of a healthy organic soil are

- Soil life
- Organic matter
- Minerals
- Air, water, and physical properties

## The Living Soil

Just as when you feed a cow well you are feeding the rumen microbes, when you feed the soil you are feeding soil microbes. Soil microbes make nutrients available to plants in similar ways that microbes in the digestive system of a cow make nutrients available for her growth and production. Keeping the microbes well fed, housed, and cared for in both soil and livestock will generate the most healthful milk possible via contented cows eating high-quality plants.

In the rhizosphere (the zone around the root hairs), an incredible diversity and abundance of microorganisms make nutrients available to plants, while plants in return provide these organism with nutrients. Most of us are familiar with the nitrogen-fixing organisms associated with clover and other legumes. They are just one example of how beneficial soil organisms are to overall soil health and productivity.

In order to thrive, the soil organisms need good drainage, adequate moisture and air, the right amounts and balance of nutrients, and the timely addition of organic matter. Crop rotations that include sod crops are particularly helpful in restoring beneficial organisms to the soil, as is the addition of compost. Compost can also increase soil biological activity on organic dairy farms. More detailed information on this is found in the companion NOFA handbook *Compost, Vermicompost, and Compost Tea* by Grace Gershuny. Dr. Elaine Ingham has done some interesting research on soil and compost organisms. Her Web site (www.soil-foodweb.com) contains information on soil microbes and their relationship with plant nutrition, soil health, and plant growth. Many useful ATTRA publications address soil organisms, compost, and soil testing.

## Organic Matter

The amount of organic matter in the soil affects almost every other aspect of soil, and is a central aspect of soil management. Organic matter should be brought to a good level and maintained carefully in order to feed soil organisms, add new ones, help build stable soil aggregates, and provide nutrients. Organic matter helps bring balance to all soil types; it improves water-holding capacity, drainage, porosity, and aeration. Soils high in organic matter have good CEC (cation-exchange capacity), allowing them to hold more magnesium, calcium, and potassium (cations). Additionally, the increased biological activity will improve the availability of other nutrients including phosphorous and nitrogen. Some methods of increasing soil organic matter and creating and maintaining healthy soils include:

- Crop rotation so that tilled land is returned regularly to a sod (grass/legume)
- Good grazing management
- Adding compost
- Minimizing tillage
- Keeping bare soil covered whenever possible by undersowing annual crops such as corn, and the use of green manures, mulch, and cover crops whenever possible
- Practices that reduce soil erosion
- Elimination of highly soluble fertilizers or synthetic pesticides

## Minerals

> By a fertile soil is meant that nature's law of return has been faithfully applied. So that it contains an adequate amount of freshly prepared humus made in the form of compost from both vegetable and animal wastes.
> —Sir Albert Howard, *An Agricultural Testament*, 1946

Either a deficiency or an excess of any mineral can lead to problems with plant growth, resulting in lower forage quality and less than ideal nutrition for livestock. Many soil nutrients can be added in the form of compost or manure, or from legumes. However, if a specific mineral is deficient, and on-farm compost or manure isn't able to bring the soil into balance, the farmer may buy off-farm amendments to add to the soil or use as foliar applications. Soil tests are an important tool to determine if soil minerals are deficient or out of balance, and are essential when adding minerals to avoid overfertilizing or creating new imbalances.

The types of amendments available and allowed for use on an organic farm include:

- Rock dusts and mined natural minerals (such as lime, rock phosphate, greensand, Sul-Po-Mag)
- On-farm produced composts, manures, or compost teas
- Off-farm manure or compost (be sure to check the source for prohibited additives)
- Commercial organic foliar and soil amendments such as fish, seaweed, and blended natural fertilizers (check the ingredients, and check with your certifier)
- Trace elements or micronutrients, but these should be added only if indicated by soil-test results, as many are toxic if applied at too high a rate (copper, zinc, and boron, for example); always check with your certifier to make sure the product is allowed for use

Nitrogen deserves a special note. It is a nutrient that most organic dairy farms can be self-sufficient in, and is usually the most cost-prohibitive nutrient to purchase in an organically approved form. In most cases, the use of legumes and careful management of manure and soil life should be sufficient to meet nitrogen requirements.

## Air, Water, and the Physical Properties of Soil

> The best-aggregated soils are those that
> have been in long-term grass production.
> —Preston Sullivan, ATTRA,[*]
> "Sustainable Soil Management"

Soils need to be able to hold enough water to get through dry spells and contain enough air so they don't become anaerobic. They should be able to resist erosion by wind or water, absorb water quickly during a rain, hold nutrients, and sustain an array of soil organisms and plant life. Are your soils naturally rocky, sandy, silty, or clayey? The basic texture of a given soil derives from the mineral of which it is made and its geologic history. Data about local soil types and the parent materials will help you understand your soils. A soils map for your area is generally available from the local Natural Resources Conservation Service office (NRCS).

Soil structure, which shows up visually in a shovelful or a handful of soil, refers to how the soil holds together in a clump (aggregation) or breaks apart. A soil with good structure and normal moisture conditions will crumble easily in your hand, but will hold together if squeezed.

Practices that improve soil structure include:

- Minimizing tillage
- Crop rotation that includes a sod
- Good grazing management
- Consideration of soil moisture conditions when using equipment on fields
- Any method that adds organic matter

## Measuring Soil Health

If we dig up healthy soil, especially in the early morning or near dusk when dew is present, an invigorating burst of pure moist earth strikes us at a most visceral level, invigorating something quite instinctual within us.
—Hubert Karreman, DVM, *Treating Dairy Cows Naturally*

Each farmer develops his or her own way of assessing soil health and monitoring changes over time. This may include tracking changes in crop yields, changes in the types of plants growing in a field, or having a lab do soil tests to decide which amendments to add. For an experienced farmer, it may also include a more intuitive ability to look at the soil, smell it, feel it, see how the plants are growing and know whether it is improving or not.

Here are some guidelines for assessing soil quality:

- Look at a soils map of the farm to understand the potential productivity of each area and soil type. Is production increasing or decreasing?
- Watch for areas of poor drainage or droughtiness.
- Watch out for areas of soil compaction. How much effort does it take to push a fence post into the ground?
- Look at soil color, smell, structure, and texture. How is it changing over time?
- See how deep plant roots grow, and how this changes over time. Is there a pan that prevents roots from going deeper?
- Watch for changes in yield or the types of plants growing in different areas.
- Observe earthworm activity by counting the number of worms or wormholes per shovelful or looking for worm castings on the soil surface. How is this changing from year to year?
- Keep track of how long it takes for a cow pie to completely disappear from the pasture, and the variety of creatures that assist this process of nutrient cycling.
- Look for plants with symptoms of deficiencies (yellowing, purple color, or spotty, for example).

- Monitor changes in yield and growth patterns in fields.
- Watch during rains to see how soils absorb water and how long they hold moisture.
- Watch for where dead plant material sits on the soil surface, and where it is more quickly incorporated into the soil.

Some considerations for soil testing include:

- Make sure the lab tests for organic matter.
- Take the samples following instructions from the lab.
- If you find a soils consultant you want to work with, choose a lab whose test he or she is familiar with.
- Use the same lab and repeat the tests every few years to track changes.
- Test for trace minerals.

ATTRA has several publications on assessing soil health, lists of alternative soil-testing labs, and other soil-related resources.

## Manure Management

> Good compost is the best way to reverse the depletion of humus on your farm. Improvements in the physical and biological realms of your soil will follow right behind an ongoing program of composting.
> —Jack Lazor, Butterworks Farm, Vermont

It's wise to consider the manure produced on an organic dairy farm as much a resource as the milk. Manure-management systems on organic farms vary widely and include composting, liquid pits, and solid-stacked and deep-litter-bedded packs. Much of the time, the preexisting facilities on the farm determine the housing and the manure system. However, increasing numbers of organic dairy farmers are shifting to some type of composting system. Many organic farmers feel they get higher forage quality from the application of composted manure than from liquid manure, particularly on farms that grow mostly or entirely perennial grass/legume pastures and hay.

When assessing a current nutrient management system or considering a new one, some questions to ask include:

- Does your current manure-management system allow collection and storage of nutrients in such a way as to be able to use them where they are most needed on the farm?
- Do you need a nutrient-management plan that is approved by a regulatory agency?
- How will you manage milk-house waste?
- How will you store and manage manure?
- Do you have access to enough bedding to make compost or a bedded pack?
- Do you have enough land to spread all the manure from your farm at the correct rates?
- Do you have enough manure to spread on all the areas that need it?
- Is the current system creating any environmental problems or nuisances such as increased fly population?

Other considerations with regard to manure include these:

- A good grazing system will provide the most even distribution of manure by grazing cows.
- Application of compost instead of raw manure to pastures can help reduce rejected forage in the next grazing.
- Healthy soil biology incorporates manure into pasture and meadow soils as rapidly as possible to minimize nutrient loss.
- In tilled soils, incorporating manure or compost as soon as possible after application, as well as seasonal timing, will minimize nutrient loss.
- A manure-storage system designed to eliminate runoff decreases nutrient loss, maximizes the quality of winter manure, and helps protect our environment.

## FEATURED FARM:
## Bragg Homestead, Sydney, Maine, Wayne and Peggy Bragg

(Excerpt reprinted with permission from an article by Jean English in the Summer 2004 issue of the *Maine Organic Farmer & Gardener.*)

Cutting his herd to the point where he could raise most of its feed and tend each cow carefully were keys to Bragg's transition to organic dairying. Transitioning was not difficult because he hadn't used synthetic or prohibited chemicals on his fields, and he had an abundance of land. He had 63 cows then, and he thought that cutting back to about 20 might be manageable. "I sold cows all through the summer when the price of milk was low. By fall I was down to 38 or 40 cows. By the following fall I was down to 31. Then I had 14 heifers calve, that brought me up to 44." He has 41 milking now, "basically a Holstein herd," and several Holstein-Jersey cross calves, as well as one Shorthorn, and this number is working well for him. He's been selling organic milk to Horizon Organic Dairy, based in Colorado, for almost two years.

Being able to get rid of problem cows eased his transition to organic, too. Instead of the expense of treating them and having the vet visit the farm, he was able to sell them and reap some income. The cows each get 3–4 pounds of cornmeal with minerals twice a day and produce an average of 50–55 pounds of milk in the summer. He feeds his cows grain containing 16 percent protein through the winter.

Bragg gives his cows aspirin for some problems. Mastitis is controlled primarily through careful attention to milking and by culling problem cows. "Milking is serious business," says Bragg. "You want your best man milking. I'm wicked fussy." When his wife or daughter help milk, they use four milkers; when he's milking alone, Bragg uses only three milkers so that he's aware of each cow's production and potential problems. He makes sure quarters don't get overmilked. "Having the machine on after the udder is empty causes stress on the udder."

three

# Plants: Crop Production and Grazing Management

Dairy cows are living, breathing, sentient creatures [that] eat grass and turn it into milk. . . . They, along with other ruminants, are here on earth to fill an extremely important ecological niche—to digest plant material that we humans cannot. We are the beneficiaries of this. . . ."
—Hubert Karreman, DVM, *Treating Dairy Cows Naturally*

## Crops

High-quality pasture and forages are essential for organic dairy production. Forages make up the greatest part of a healthy, balanced dairy ration, and forage quality has a large effect on farm profitability. Offering cows quality pastures during the grazing season, and quality stored forages in the winter, will allow the ideal production of milk while keeping the livestock healthy and long-lived.

Depending on soil types and the crops grown on the farm, pastures and hayfields may be permanent perennial fields, or they may rotate with some annual crops. Farmers who have access to the equipment, labor, and suitable land can rotate grain, corn silage, or other annual crops with grass/legume sod. Each farm will have to assess the feasibility of growing annual crops according to their farm-family goal and the economics of production. Grass/legume sod, whether used in rotation with annuals or as the main forage crop, offers significant benefits in building and maintaining soil fertility.

Perennial grass/legume sods:

- Add organic matter to the soil
- Add nitrogen
- Improve soil structure

- Decrease runoff and increase water infiltration
- In rotation, can reduce weeds and pest pressure
- Provide habitat for beneficial insects
- Produce quality forage when managed well and fertilized correctly

Homegrown forages offer the cheapest form of nutrients available to feed dairy cows. In organic dairy production, protein is generally the most expensive purchased concentrate (often in the form of soybeans or another leguminous grain). Energy concentrates tend to be less expensive (corn, barley). Thus it pays to grow high-protein forages, but with attention to the quality, not just the quantity, of protein in the plants. Too much protein in the ration, particularly highly soluble or "nonprotein nitrogen," can cause health problems. Low-quality forages due to late harvest or poor weather conditions are another potential feed-quality problem, resulting in high-fiber forage that is not easily digested.

A farm that produces high-quality pasture during the grazing season and has quality stored forages to feed in the winter can formulate a ration that's lower in concentrates, and consequently has fewer health problems in the herd. For farmers trying to shift toward low- or no-grain dairy farming, high forage quality is essential.

## Stored Forage

The quality of stored forages is improved by:

- Improving soil health with attention to relative and total amounts of minerals
- Selecting locally adapted species of high-quality grasses
- Increasing the legume content
- Harvesting at the right stage of maturity
- Harvesting when weather is favorable
- Using an appropriate harvest method
- Packing and storing silage or haylage to minimize spoilage

Legumes provide high-protein feed—generally the most expensive feed to buy organically—while putting nitrogen into the soil (nitrogen is the most expensive fertilizer to find in an organically approved form). By optimizing the use of the farm's manure and making good use of legumes in forage mixes and crop rotation, soil health and forage quality can both improve. Many legumes, unlike most grasses, can be added to a field by frost seeding or overseeding just 3 or 4 pounds of clover seed per acre. Over time, grasses tend to outcompete legumes by shading and crowding them out, particularly if there are soil imbalances. Good pasture management will help maintain the best balance of legumes and grasses in a pasture.

**Well-fed and happy. On most organic dairy farms, grass/legume meadows and pastures make up the central portion of the cropping system.**

To improve the success of a frost seeding:

- Use heavy fall grazing to control grass growth and create more open soil.
- Test soils and fix imbalances that hamper legume growth. Pay particular attention to lime, potassium, boron, and phosphorus.
- Don't apply nitrogen fertilizer or liquid manure, as this will encourage grass growth.
- Seed in late winter or early spring.
- Graze in the spring to keep the grass from getting too tall too early and shading clover.

Before plowing and reseeding a grass/legume pasture or hayfield, be sure that you haven't overlooked other options for renovating and improving the field. More economical ways to improve the area may include frost-seeding clovers, no-till drilling-in seed of improved grass species, changing the grazing management, using an aerator or ripping with a chisel plow to improve aeration, or adding compost or some other amendment to improve soil health. When reseeding is necessary, either as part of a rotation with annual crops or for some other reason, choosing the right species is important. Selection of the right species, variety, or mix of species involves a consideration of soil pH, drainage, fertility, climate, palatability of the crop, weed pressure, the harvest method, and the length of time the stand is needed.

When buying any type of seed, be sure that it is not genetically engineered or treated with a fungicide. The use of GMO or fungicide-treated seeds is not allowed on organic farms. GMO seeds are often not clearly labeled, and it is up to the farmer to find out. Organically certified farmers need to buy organic seed whenever it is commercially available.

## Forage Species

### Annual Crops

**Brassicas,** which include forage varieties of kale and turnips, are high-yielding, highly digestible crops that can be used for summer or fall grazing.

Grazing animals must be introduced slowly to brassicas, and they should only be used to provide a portion of their feed to prevent health problems. Brassicas are not drought tolerant and do best in soils with moderate to high fertility. They may be grown in a mix with other annuals or drilled into an existing grass or legume crop.

**Grains** such as oats, annual ryegrass, and triticale can be grazed or harvested as forage or grain. They are generally a high-energy crop, either as a forage or grain. Their relatively short growing season makes it possible to mix them in with other crops in a rotation or under-seed them with a legume. They may be grown as a mix of several species with a legume for either grazing or harvest.

**Corn** is a high-yielding warm-season grass that requires a high level of soil fertility. It can be harvested as a grain crop, as silage, or it can be grazed. Organic corn will require a good crop rotation to maintain soil fertility, and a weed-management system based on prevention through rotation and cultivation.

**Peas** and **beans** (which are legumes), including soybeans, are generally grown as a grain crop, but can also be grown as a silage crop or green manure. They are a high-protein feed. Peas may also be grown in a mix with an annual grain and either harvested or grazed.

**Warm-season grasses** such as BMR sorghum or millet are fast-growing, high-yielding crops that can be harvested as silage or grazed. In some situations, sorghum may have a risk of prussic acid poisoning. Millet does not carry this same risk.

### Legumes

**Alfalfa** is a higher-yielding legume than many of the other legumes. It is often grown with a grass such as timothy or orchard grass and harvested. It has a deep taproot so it can tolerate dry conditions. The length of time a stand persists depends on climate, harvest management, soil fertility, and soil conditions. It tends not to persist very long in poorly drained acidic soils, or under intensive grazing.

**Alsike clover** is a short-lived perennial clover that will do well in poorly drained soils. It has a growth habit similar to red clover, with less rigid stems. Unlike red and white clover, it doesn't tend to establish well from frost seeding. It is suitable for grazing or haying.

**Bird's-foot trefoil** is slow to establish, but does well in acidic, moist soils. Unlike the other legumes, trefoil does not cause bloat. It is similarly high-quality as alfalfa, but lower yielding. There are varieties selected for grazing or haying. It is good for late summer or fall stockpiled grazing.

**Hairy vetch** is a short-lived perennial that can grow in a range of soil conditions. It has a vining growth habit and can be either grazed or harvested as part of a hay-crop mix. If grazing this plant, it does best if it is not grazed short.

**Red clover** is a short-lived clover, generally persisting for two or three years if it isn't frost-seeded or allowed to go to seed. It is often grown with hay-type grasses to improve forage quality and provide nitrogen, or as a green manure or cover crop. It tends not to persist under grazing as well as white clover, but with occasional frost seeding into a grass mix, it can produce high yields of hay, silage, or pasture.

**White clover** is a shallow-rooted legume that is well adapted to grazing. It can be frost-seeded into a pasture to improve forage quality and spreads naturally under good grazing management. There are shorter and taller growing varieties available. White clover has a hard-coated seed that frost-seeds well, and under certain conditions will sit dormant for many years before germinating. It mixes well with cool-season perennial grasses in a pasture mix.

### Perennial Grasses

**Bromegrass** is a high-yielding grass that has an above-average tolerance to drought. It tends to need a longer regrowth period after cutting or grazing, so it requires careful management, particularly in pastures. It is slow to establish, doesn't do well in low-fertility soils, and can be difficult to keep in a pasture mix.

**Kentucky bluegrass** is a low-growing, sod-forming grass found in pastures throughout the United States. It is well adapted to grazing and grows well with white clover. It is very winter hardy and grows well in the spring and fall, but production drops during hot or dry weather.

**Orchard grass** is a tall-growing bunchgrass that does well under managed grazing or haying. It has fair tolerance of drought and cold winters and is relatively easy to establish. Over time, stands can thin out, leaving bunches of grass with bare soil or other species between. Frequent grazing or mowing helps maintain its quality, and after the seed heads have developed and been removed by harvest or haying, growth for the rest of the year is all high-quality leaves. It grows well with red clover or alfalfa, and the later-maturing varieties are often easier to manage in a mixed alfalfa/grass field.

**Perennial ryegrass** is a high-yielding nutritious forage suited to either grazing or hay/silage production. Some varieties are less winter hardy than many of the other cool-season grasses. It is a high-energy grass that does well under grazing. Owing to the marginal cold hardiness of some varieties in more northern areas, it works best in a mix with other grass and clover species.

**Reed canary grass** is a tall-growing, sod-forming, high-yielding grass. It is slower to establish than some other grass species, but has high winter hardiness, does well in droughty conditions, and grows well in wet areas where other grass species don't do well. This species can grow even in the "summer slump" when other species are less productive. Manage it carefully to avoid it becoming overmature, which reduces quality and palatability. Some varieties have low palatability due to the presence of alkaloids, but there are newer lower-alkaloid varieties available.

**Tall fescue** is a deep-rooted grass that can be grazed or harvested for hay or silage. It is one of the most drought-tolerant cool-season perennial grasses. This grass is much less palatable than other cool-season perennial grass species, so in areas that do not have extended hot, dry periods of weather it may be rejected by grazing cattle that will prefer to graze the other grass and legume species. Some tall fescue varieties are also high in endophytes, which can contribute to low palatability and have some adverse affects on livestock.

**Timothy** is a shallow-rooted bunchgrass that produces high-quality hay or silage. It is often grown with red clover or alfalfa for hay. It doesn't persist well under intensive grazing, although some newer varieties are exhibiting more suitability. Different varieties are available that mature earlier or later, depending on which legume type it is grown with.

Many dairy farms have land that is used for both grazing and haying, depending on the time of year. Many of the cool-season perennial grasses will do well under either good grazing management or hay management. Areas that are grazed more will tend to have plants that are better adapted for grazing, such as white clover, Kentucky bluegrass, perennial ryegrass, and other sod-forming grasses. Areas that are primarily mechanically harvested will tend to have fewer grazing-adapted species such as alfalfa, red clover, timothy, and bromegrass.

The stage of maturity of harvest depends on production goals and the type of plant. Cool-season perennial grasses, often grown with legumes, are generally harvested at boot stage or early bud. However, alfalfa is often harvested in early bloom to increase crop persistence. Annual cereal-grain crops may be harvested early for the production of high-quality forage, or they may be left to mature for grain harvest.

Method of harvest depends on equipment and capital available and which crops are grown, as well as on storage facilities and local weather conditions. Some farmers like to have a high-energy forage such as corn silage to feed along with high-protein grass silage in winter or with pasture during the grazing season. Many farms continue to make at least some dry hay to feed to some or all of the livestock. However, other farmers have shifted entirely to silage or baleage production owing to ease of harvest and storage along with local weather and drying conditions that impede the production of high-quality dry hay.

Dairy farms that have suitable soils and the right equipment may also benefit from growing annual crops for silage or grains. Some organic dairy farms grow their own corn, and increasing numbers are growing some small grains. As the price of supplemental protein has risen recently, there is also increasing interest in growing soybeans, peas, and other high-protein crops.

Summer annuals may be useful as both grazing and harvested crops. During the "summer slump," when cool-season grass pastures are regrowing very slowly, annual crops such as millet and BMR sorghum can provide quality pasture while allowing the permanent pastures a chance to regrow for fall grazing.

Crop rotation is an essential part of building and maintaining a healthy soil in fields where annual crops are grown. The practice of rotating crops can increase yields, decrease disease, pest, and weed pressure, and improve soil conditions. Legumes in a rotation fix nitrogen, and accomplish it best in a long-term rotation that includes a grass/legume sod. Including cover crops in the rotation is a way to reduce soil erosion, build soil, add nutrients and organic matter, and disrupt pest life cycles. This can be particularly useful for winter soils that would otherwise be bare. Cover crops can go in after the crop has been harvested, or be undersown with the main crop.

### Assessing Forage Quality

> The lab test may quickly identify gross excesses or deficiencies in the feed and thus enable you to make adjustments before problems occur. It does not hurt to have two opinions . . . one from the lab and one from the consumers, your animals.
>
> —Richard J. Holliday, DVM,
> *Fundamentals of Holistic Animal Health*

Besides knowing how to produce high-quality forages, it is important to understand how to assess the nutritional value of forages to make decisions on the appropriate types and amounts of concentrates and other supplements to feed.

Some methods of assessing forage quality include the following:

- Results of lab forage testing
- Visual assessment of color, amount of stem, amount of leaf
- Smell: Is it moldy? Dusty?
- Watching which forages the animals prefer
- Livestock performance: growth, reproduction, milk production, animal health

The lab forage test is helpful in determining feed quality and deciding what types and amounts of supplemental feeds to use. However, it doesn't tell you everything. In assessing forage quality, consider these:

- Digestibility—higher-fiber forages take longer to digest, and the rate of passage through the cow's digestive system will affect how much she will eat. Digestive problems can develop when fiber is lacking or chopped too short.

- Energy—younger plants generally contain more energy, as they contain more digestible fiber. Better soil fertility and mineral balance can also improve forage energy content.
- Protein—the amount of protein and the amount of nonprotein nitrogen is influenced by soil fertility, plant species, and the stage of maturity.
- Minerals—soil imbalances, combined with some plant species, can result in forages with mineral imbalances that can create health problems.
- Vitamins—a cow with access to sunlight and fresh high-quality forages will be able to get or make more vitamins. The vitamin content of stored forages can decline over time.

## Pasture Management

Pasture is required by the organic standards, and also provides low-cost, high-quality forages and a healthy, low-stress environment for livestock. The methods of pasture management on organic dairy farms include management-intensive grazing (MIG) and holistic-planned grazing (HM, or holistic management) where cows are moved to a new, high-quality pasture twice or more every twenty-four hours; less-intensive rotational systems; and large extensively grazed pastures.

When compared to MIG or HM, the rotational systems and large-pasture system require less daily management of moving fence and water tubs. However, the extensive or rotational system will require more acreage to supply the same amount of dry matter, and it may require more clipping or occasional pasture renovation over time owing to overgrazing damage. Rotational or extensive grazing will also provide a more variable quantity and quality of feed throughout the grazing season.

The type of pasture management used should be determined by the farm's overall goal and the production objective for the livestock. In this handbook, the focus will be on MIG, as that type of management will provide high-quality pasture and good animal performance.

## Grazing

> It is no accident that good pastures have perhaps 40 to 50 species. . . . Nor is it an accident that fully half the weeds in the USDA index are also listed in manuals on medicinal plants.
> —Gearld Fry, *Reproduction and Animal Health*

When animals go into a pasture, they aren't just eating. They are trampling weeds and dead plants into the soil, which adds organic matter. They are selecting the plants that they want to eat and spreading manure on the pasture. Well-managed perennial pastures provide livestock with a "salad" of many types of plants. There may be many different species of plants in a single pasture. This biodiversity provides the cow with satisfying and beneficial choice and means that even in a drought or a very wet year, something will grow.

### Benefits of Grazing for an Organic Farm

Pastures provide low-cost, high-protein feed—particularly helpful for organic dairy farmers, for whom the costliest supplemental feed is protein. When grazing high-quality pastures, organic dairy farmers can switch to a low-protein, high-energy feed like corn, barley, or molasses and not have to pay for higher-priced organic soybeans and other proteins.

Good grazing management can improve livestock health, resulting in reduced cull rates, longer-lived animals, lower vet bills, and additional income from sales of heifers and cows. Good grazing practices can improve pasture quality and yield. They can convert weedy or brushy pastures, where animals have to search to find good-quality forage, into highly productive pastures that can feed more animals high-quality forage, produced and harvested at a low cost.

Farms that direct-market and are both organic and grass-based have a market niche in satisfying the consumer desire for organic, grass-fed food, as well as the public interest in aesthetic, humane, and environmentally beneficial agriculture. (See chapters 5 and 8.)

## Potential Challenges

When not managed correctly, pastures can become a main source of parasite infection, particularly for young stock and certain types of livestock (sheep and goats). A good understanding of the life cycle of parasites and how to design a grazing-management system to minimize risk of infection is needed to reduce or avoid reliance on chemical dewormers. Knowledge of alternative parasite-management methods can also be helpful. (See the internal-parasite management section in chapter 4.)

An incorrect supplemental feeding program can result in too much supplemental protein being fed while animals are on pasture. Poorly managed grazing can cause inadequate intake of dry matter. These are two common grazing-related management mistakes that will result in loss of body condition, animals not breeding back, lower milk production, and slower growth. Development of a supplemental feeding system appropriate to the pasture quality and quantity is an essential step in setting up a successful grazing system.

Good MIG will favor desirable pasture-plant species, reduce weed problems, and increase the quantity of pasture dry matter produced while improving the nutritional quality of the feed. A high-quality pasture, particularly with good care of the soils, will produce livestock feed of the highest nutritional value and vitality, so that animals are healthy, and the meat, milk, and manure produced is of the highest possible benefit to you and your farm.

A good-quality MIG pasture will be a mix of many plant species, with no bare soil showing, with uniformly distributed cow pies from the most recent grazing. The pasture will have patches that did not get grazed closely during the last grazing, since cows don't like to eat the grass right next to their manure.

In an extensively or continuously grazed large pasture that is not rotated, there are more likely to be patches of bare soil, less diversity of cool-season grasses, clovers, and other legumes, and an increase in types and numbers of weeds. Patches will appear that never grow very tall, and clover and other legumes may be completely absent. There may be a buildup of dead plant material or thatch on the soil surface, and cow pies may not decompose quickly.

Pastures do not suddenly become poor in quality. They decline gradually. Knowledge of soil fertility and a good understanding of grazing-management techniques along with trained observational skills are needed to maintain this vital part of the organic system.

**Pasture management done well results in pasture with a mix of forages that includes legumes, which offer a source of low-cost, high-protein feed. There's also no bare soil, cow pies are uniformly distributed, and there's easily accessible water.**

### Management-Intensive Grazing

The basis of good grazing management is that pasture plants need time to rest after each grazing in order to photosynthesize and replenish energy stored in their roots. Continuously grazing animals in the same pasture, or returning them to a pasture before it is fully regrown, does not give the plants time to recover. This pattern results in weak plants that may stop growing and die. These weakened plants will not compete well with weed species, and won't hold soil well, resulting in bare soil and erosion. Some grasses and clovers will survive by staying very short, never growing tall enough for livestock to easily graze, while livestock will reject other areas that will soon grow up into weeds, brush, or small trees.

Under MIG, many dairy farmers will give animals a fresh pasture after each milking. Livestock may return to the pasture when it has fully recovered by regrowing to 6–10 inches in height. This regrowth period may be as short as fourteen days when the plants are growing rapidly, but it may be forty days or longer later in the summer. Irrigation may be needed in some areas to allow regrowth.

It is important to move animals frequently so that each paddock is not grazed for too many consecutive days in a row. Grazing periods of half a day to one day will result in much higher pasture quality and livestock performance than grazing periods of several days to a week.

Another important part of management-intensive grazing is that when pasture growth slows, either the total number of acres being grazed needs to be increased, or the paddock sizes need to be decreased (and more supplemental feed must be fed) to slow the rotation. If the rotation is not slowed to correspond with slowing plant-growth rates, plants will not be getting enough rest, and dry-matter intake by animals will drop, resulting in poor pasture and animal performance. Timing the first cut of hay early enough to allow some areas to grow back tall enough for grazing is the easiest way to have good quality pasture available when plant growth slows in summer.

For farmers planning to adopt MIG, it can be helpful to attend a grazing school to learn how to set up a system, estimate dry matter, and gain other important grazing skills that this handbook doesn't address in detail. Look at some of the grazing books, magazines, and resources listed in the resource section.

## Setting Up a Grazing System

### *Resources*

- Attend pasture walks, visit other grass-based farms, and learn from their experiences.
- Contact your local Natural Resources Conservation Service (NRCS) office and ask about programs that help with the cost of fencing, water, and creating travel lanes.
- Locate some of the grazing publications, articles, Web sites, books, and organizations listed in this handbook.

### *Pasture Design*

- Put travel lanes on high dry ground—you will probably need to do some improvement and maintenance on muddy wet areas.
- Try to provide water in each paddock so animals do not have to walk to find it and drop manure where it is not needed.
- Whenever possible, try to put fast-growing areas in one paddock and slow growing areas in another.
- Consider topography; put south-facing slopes in one paddock and north-facing slopes in another.
- Put gates in the corner of the pasture closest to the barn.
- Minimize shady loafing areas where animals will tend to nap and drop all their manure during weather when they don't need the shade.
- During hot weather, strategies farmers use include grazing "shade pastures"—set aside for the hottest summer days, keeping cows in a well-ventilated barn during the hottest days and grazing at night, moving cows more frequently to high-quality pasture, using sprinklers, and even some portable shade systems. How cows are managed during the hottest time of the year depends on the temperature, humidity, facilities available (some barns are hotter inside than it gets outdoors), amount of shade available on the farm, breed and adaptation of the cows, and the farmer's personal beliefs.

### *Fence and Water Systems*

- To start with, try to be as flexible as possible with your fence and water system. You may change your setup a few times, and you may need to have flexible paddock sizes as your herd size changes and your pasture productivity increases. You can use temporary internal fences (step-in posts with poly wire or stranded aluminum wire) and, to begin with, put up permanent fencing only around the perimeter of the pasture system.

- Get a good quality energizer and be sure it is well grounded, with lightning protection.
- Build perimeter fencing that can conduct electricity with minimal resistance. High-tensile fencing makes a permanent fence that is effective, safe, and low maintenance.
- Use high-quality poly wire and posts. Learn how to make good connections for each type of wire.
- Provide water in each paddock whenever possible.

#### *Grazing Guidelines*

- Try to walk your pastures each week and record how tall each pasture is (to determine how much dry matter is available for the animals to graze).
- Let animals back into an area only after it has fully regrown.
- Limit grazing periods to three days—twelve to twenty-four hours is better.
- Move animals frequently to help increase dry-matter intake and improve pasture quality faster.
- Avoid a set rotation. Graze according to plant-growth rates. If one pasture grows faster than the others, graze it more often. If you have a pasture that grows slowly, graze other areas and let the plants grow back.
- Lock animals in each paddock so they can't wander back to the barn. This is one of the easiest ways to manage soil fertility in pastures.
- When strip grazing, consider using a back fence to prevent "back grazing," so that animals don't overgraze favorite plants.
- When possible, use a follow-up or clean-up grazing group. On a dairy farm the milking herd can graze the paddocks first (and get the higher-quality forage), followed by a group of dry cows that graze the rejected forage and "clean up" the pasture. If you don't

use a follow-up group, you may want to clip some pastures to improve palatability and quality.

- Don't let grasses get too tall and shade out white clover. Overmature plants are lower-quality feed, and too much shading may decrease the plant density in the pasture.

## FEATURED FARM: B-A-Blessing Farm, Whitesville, New York, Tammy and John Stolzfus

(From an article by Tammy and John Stolzfus. Excerpt reprinted with permission from *NODPA Notes* 4, no. 1.)

Although our former milk handler claimed we couldn't do it because of high somatic-cell-count (SCC) problems, we have found that the longer we are organic, the lower the cell count has become. Dr. Ed Schaeffer has a product called Mastoblast®, which has helped with the cell count. However, John has found that raw garlic is very successful in treating mastitis. We give an infected cow 3 pill guns full, twice a day for 3 days or until the mastitis clears up.

One of the things we have learned since going organic is that calves get a much better start in life if you leave them on their mothers for at least 3 weeks. The only disadvantage to it is a very noisy barn when you take them away from their mothers.

Another thing that we are trying to get away from is feeding a lot of grain, as we have to purchase all of it. We feel it is much better for the cows' overall health if they do not eat a lot of grain.

Currently, the hay we are feeding is dry round bales and baleage. John's goal is to have a herd average of 17,000 to 18,000 pounds of milk with a butterfat of 4.0 and protein at 3.2, while feeding 2–5 pounds of cornmeal and an all-baleage mixture of oatlage and mixed grasses with clover.

Our milking herd started out as an all-Holstein herd but we are now crossbreeding Holsteins and Normandes. There are also a couple of Brown Swiss mixed in the herd. We have a total of 79 cows plus 42 head of young stock and bred heifers. It seems to be an ongoing challenge to get cows bred so that there are five to six freshening every month. During the winter there is a period of time when there are no fresh cows. This causes the bookkeeper (Tammy) to have some tense moments when it comes to paying the bills.

As much as possible, John keeps the cows on rotational grazing lots with the animals going onto a new lot every 12 hours or oftener. We have water available in about half of the lots and plan to make it available to more lots each year.

We have also started to compost manure and have found that it is fairly simple to turn it over with the tractor bucket about every three weeks.

---

four

# Livestock

## Selection

Selecting and finding livestock with appropriate genetics, well suited to high-forage diets and organic management, is important. Having the right animals will make management on an organic farm a lot easier.

In the long run, most organic dairy farms try to select cows that have good fertility and resistance to disease and are long-lived. These cows should have the ability to do well on a high-forage ration, be adaptable, and have good production over their lifetime. They may not be the cows that produce the largest volume of milk in their first lactation, but will be cows with good lifetime production and minimal health problems.

While some buy an organic herd when starting a new organic dairy farm, others will transition an existing herd from conventional to organic management. During the twelve-month herd transition, culling choices will need to be made to eliminate cows that do poorly under the new organic-management system. The ideal situation is one where there are enough replacements coming into the herd that animals can be sold or culled based on keeping the best-suited animals in the herd. As selection is done, and soil and herd health improves, the hope is that animal longevity will increase to allow more selection choices.

When choosing which bull to breed to, or which cows to buy, it is essential to be clear about what your breeding-program goals are. Otherwise, some salesperson will be happy to provide you with genetics that meets *his* or *her* goals.

What traits do you want to improve in your herd?

- Calving interval and reproductive performance?
- Longevity?
- Rate of gain and age of puberty?
- Grazing ability and performance on high-forage diet?

- Milk solids and total milk production?
- Calving ease?
- Mothering ability?
- Good feet, legs, and overall body type?
- Udder and teat traits?
- Behavior and disposition?
- Adaptation to the local environment?
- Parasite resistance and strength of immune system?

And here are some other considerations:

- Why are you choosing this breed?
- What other breeds have traits you like or don't like?
- Will you use a bull or artificial insemination (AI)?

Much of the success of AI depends on the herd manager's ability to detect cows in heat and breed them in a timely fashion. Hormones to cycle cows are not allowed in organic production. Whether bull breeding or AI is the method chosen, selecting the right genetics to fit the farm-management system is important.

## Nutrition

Nutrition is an extraordinarily complicated chemistry. In addition to well-known requirements for protein and energy, there are requirements for minerals and trace minerals that extend beyond our understanding.
—Dave Hoke, DVM, *A New Troubleshooter's Guide to Dairy Cows*

The type of feeding system, along with the selection and quantity of forages, concentrates, and supplements, make up the nutrition program. In this mix, the quality of stored and grazed forages is most important.

Feed rations for ruminants consist of forages including hay, pasture, and silage as well as grain (concentrates) and supplemental minerals. Vitamins can also be included. Digestion of fiber and other materials in the rumen is what makes many of the nutrients available to the animal. Therefore, much of the feeding of cows can really be considered feeding the microbes that inhabit the rumen, which in turn will feed the cow.

Managing the quality of stored forage through timing of cutting, plant-species selection, soil fertility, and storage methods will create a foundation of quality forages leading to healthier, more productive animals. Improving the grazing management on the farm to increase total dry-matter intake of higher-quality forages during the grazing season will lower costs and improve livestock performance during the grazing season.

Ration balancing means determining the best mix and quantities of forages and grains to feed based on the nutrient requirements of animals and the quality of available forages. Nutrient requirements of livestock vary with age, production level, and pregnancy as well as according to breed, livestock selections, and how the animals have been raised and cared for. Rations can be balanced based on optimum performance, least cost, and type of feeds available or in light of overall production and farm goals. There are ration-balancing computer programs that you can purchase or access by working with a nutritionist. Rations can also be balanced by hand, using a calculator and information from charts of nutrient needs and feed quality and adding extra supplements (vitamins, minerals, kelp). For more information on this, refer to one of the resources listed in the nutrition section of the resource list.

## Digestion

> The ruminant digestive system depends upon continual inoculation of microorganisms from the habitat, beginning at birth, to complete the digestive process.
> —Dave Hoke, DVM, *A New Troubleshooter's Guide to Dairy Cows*

Cows, sheep, and goats are ruminants, which means they have a unique digestive system that includes four stomachs. This digestive system allows them to eat and digest large amounts of coarse, fibrous feeds such as silage, pasture plants, and hay. The rumen is the first stomach, and a cow's may exceed 35 gallons in size. Here, a large number and variety of microbes break down forages into substances that can be used by the cow for the production of milk, and for maintenance and growth.

These organisms in the cow help feed her in much the same way as soil microorganisms support plant nutrition. Keep the well-being of these microbes in mind when designing a ration or changing a feeding program. It will help keep the microbial community healthy and the cow well fed.

To keep the rumen and rumen organisms functioning well:

- Feed a high-forage diet.
- Do not overfeed grain.
- Allow access to free-choice trace-mineral salt, minerals, or kelp.
- Include a mineral supplement in the grain.
- Make sure livestock have access to sufficient clean water.
- Be sure cows are not eating too much protein.
- Avoid abrupt changes in feed types and amounts.

In fine-tuning the ration it is essential to constantly monitor animals *and* feed. This can involve forage testing, watching the texture of the manure, monitoring milk production and growth of young stock, and body-condition scoring. Body-condition scoring (BCS) can be a useful tool, particularly in keeping track of the energy balance of the ration. BCS is a way to assess a cow's fat reserves, and is most useful when comparing the different stages of her yearly cycle. Overly fat cows tend to have more cases of reproductive and metabolic disorders. Cows that show an excessively large change in body condition can suffer more metabolic disorders. Excessively thin cows, particularly at certain stages of reproduction and lactation, are also more prone to problems. There is an excellent discussion of BCS in Hokes' *A New Troubleshooter's Guide to Dairy Cows.*

### Concentrates

The choice of how much grain to feed, production goals of milk per lactation, and how long to keep individual cows in the herd are decisions that must be made according to each farm's management system and long-term goals. A diet high in grain stimulates higher production in the short term, but it must be balanced with long-term production over several lactations. A regimen of feeding large amounts of grain, especially if it involves less-frequent feedings of large amounts, risks rumen acidosis, which leads to a whole host of problems.

Acidosis can be reduced by decreasing the total amount of grain fed per day, spreading out the grain feeding so that a smaller amount is fed at each feeding, and increasing the amount of long-fiber hay, particularly when cows are grazing high-quality pasture. Testing for milk urea nitrogen (MUN) and blood urea nitrogen (BUN) measures the amount of nitrogen being excreted and can be used to determine if the amount of protein being fed is balanced. The smell of manure and changes in its texture are also clues to changes in the ration and cows' response to them.

### Non-Grazing Season Considerations

Here are some questions to ask when your cows are not out to pasture:

- How could you improve the quality of stored forage?
- Is there adequate space for all the cows at the feeder?
- Are cows getting adequate dry-matter intake (DMI)?
- How can adjustments be made to concentrates and supplements for better performance and lower cost?
- Is the water supply accessible and clean?

When stored forage quality is poor, more supplemental feed, including concentrates, will be needed. Forage analysis can be helpful in determining nutrient content and balancing the ration to meet the needs of the cow efficiently. Forage testing will not only give early warning of poor-quality forages, it can help reduce feed costs and animal health problems due to overfeeding by allowing a better-balanced ration. Cow genetics and breed type also strongly influence performance.

### Grazing-Season Feed Considerations

When the cows are out to pasture, these questions can help guide your management:

- What could you do to improve pasture quality and quantity for the cows?
- Do you have enough land for grazing?
- How can you adjust your concentrates and supplements for better performance and lower cost?

- Are the cows getting enough DMI?
- Are they getting too much protein?
- How far do they have to walk to water?

The two most common nutritional pitfalls during the grazing season are feeding too high a protein concentrate and inadequate DMI due to pastures being too short.

Under continuous grazing, or when using a rotational system with a few large paddocks, there may be adequate high-quality forage in the pasture during the early part of the grazing season, but as the season progresses, both quality and quantity available to the cows will decline. This can result in much lower total DMI and cows that, if not provided with additional forages in the barn, will perform poorly. Note that even under good management-intensive grazing, there are seasonal variations in pasture nutrition quality, and the additional concentrates and supplements should be modified as needed.

### Nutrients

Fiber is necessary for the mechanical functioning of a healthy rumen and digestive system. The length of the fiber is important, as well as the total amount and quality. Digestion of fiber also provides an essential source of energy, both for the cow and for rumen microbes. Cows eating enough good-quality, adequate-length fiber will have good cud-chewing activity and lower incidence of displaced abomasums or acidosis. Cows with a high intake of poor-quality fiber (low digestibility) pass the material from their rumen more slowly. This means that they fill their rumen with fiber that takes so long to digest that they cannot meet their nutritional needs.

Protein is a necessary component of body tissue and milk production. Proteins are made up of many different types of amino acids, all of which contain nitrogen along with other elements. Animals need to get some "essential" amino acids in their feeds, and they are able to make others. Feed testing will tell you the total "crude" protein, but it doesn't show protein quality, or which types of amino acids are present or lacking. In general, providing protein from a variety of plants that are grown on well-balanced (not over- or under-fertilized) soil will supply the healthiest balance of quality protein. Forage is the primary protein source for ruminants, with legumes being the key component. When forage isn't adequate to provide all the needed protein, grains such as soybeans, peas, or oilseeds like flax are generally added to the diet. The high cost of protein concentrates is an incentive for organic producers to grow and store high-quality protein forages.

Energy powers the metabolism and is needed by both the cow and rumen organisms. Starch and sugars in stored forages are a source of energy in an animal's ration. However, most organic dairy cows do best when fed some corn or another energy source such as barley, oats, molasses, or corn silage. Feeding oils or fats from oilseeds is another energy source. Growing high-energy forages or supplementing with a high-energy grain is particularly important for dairy cows when the primary forage source is lush pasture, which is high in soluble proteins. However, excess energy feeding can result in over-conditioned cows.

Most dairy farmers feed supplemental minerals (salt, kelp, trace minerals) in the grain, or in a mineral lick. Increasing numbers of organic dairy farmers use kelp, and when feeding minerals, many prefer to use loose mix instead of blocks. Mineral deficiencies can result in depressed immune function, increased susceptibility to parasites, increased metabolic disorders, and other problems. Overfeeding a particular mineral can also result in problems, so testing forages and soils and designing a mineral supplement that matches the farm is important. A section in Karreman's *Treating Dairy Cows Naturally* describes minerals and vitamins as they relate to health issues.

It is important to note that the trace minerals are essential, but toxic in larger amounts. The total amount of each is important, but consideration must also be given to the form and the ratio of each element to every other element.

Be aware that many commercially available mineral lick blocks as well as some loose mixes contain ingredients that are not allowed for use in organic production. These prohibited ingredients include mineral oil, nonorganic grain or "roughage," as well as artificial flavors and artificial colors. Read the labels carefully and check with your certifier.

### Special Considerations with Zero-Grain Feeding

There is increasing interest in producing milk and meat from animals fed little or no grain, owing to market prices for milk, and the cost of grain.

However, transitioning to no-grain feeding is challenging, and if done too fast, with the wrong genetics and anything less than the best-quality forages, it can result in poor animal performance and health. A poorly planned transition to no grain can lead to cows losing condition, not breeding back, and significant drops in their milk production.

If a farm is interested in the goal of zero-grain feeding, it is important to be clear about how that fits into the overall farm and family goal.

- Will the decrease in total milk production still allow enough cash flow to cover farm and labor costs?
- Is the quality of the stored forage and pastures excellent and consistent?
- Are the manager's grazing and feeding skills high enough?
- Is there a system to allow supplementation with enough minerals?
- Is there enough market demand, and is the price for products high enough?
- Are the animals suited to a no-grain system?

Moving toward zero-grain feeding begins by rearing grain-free young stock. Selection of suitable breeds or bloodlines is also necessary, as is having high-quality stored forages and an excellent understanding of, and skill with, grazing management.

A clear understanding of the economic incentives (or lack of them) to cut out the grain is important, since the only farms at this time being paid a premium for no-grain feeding are those with a local value-added market. If the decision to stop feeding grain is made, allow plenty of time and flexibility to feed grain as needed during the transition to no grain.

### Seasonal Milking

Seasonal dairy farming, where all the cows are dry for two months or longer, is an option only if your market will allow it. Small-scale dairy farms that make an aged cheese may choose this option, as the cheese can be sold year-round, even if there is no milk supply for a period each year. For a grass-based dairy, this can allow the production of milk during the grazing season, when feed costs are low. Some farms using this seasonal timing are also taking advantage of the high nutritional content of the milk during the grazing season to sell "high-CLA" or "high-omega-3" dairy products. Other farms dry the cows off during the summer season to allow them to focus on producing quality forages or other crops. Before choosing this management option, be sure to consider the challenges as well as the potential benefits. The challenges for seasonal dairy farmers include:

- Maintaining a twelve-month calving interval so cows breed back and calve in a narrow window;
- Finding a market or milk buyer tolerant of seasonal variations in milk supply;
- Managing a large number of cows calving, and many calves to care for at once;
- Managing a whole herd of late-lactation cows with potential increase in SCC (somatic-cell count).

### FEATURED FARM:
### Chase Hill Dairy Farm, Warwick, Massachusetts, Mark and Jeanette Fellows

(Excerpt reprinted by permission from an article by Jonathan von Ranson that appeared in the *NOFA/Mass News* June/July 2003 and *NODPA Notes* 3 no. 4.)

Chase Hill Farm is the only organic dairy in Massachusetts [at the time this article was written], and it is making a decent living. It is one of the six dairies in the "Our Family Farms" group, which through good marketing and a favorable arrangement with a processor is able to pay farmers a premium for their milk. And Jeannette has developed a cheese business with outlets at farmers' markets, farm stands, and food markets.

Early on, the Fellows attended a UVM dairy conference and took home a one-page handout about rotational grazing that seemed to make sense. Mark began moving his herd daily to new pasture, learning about pulsed grazing and its benefits to the land and the cows.

He decided to give up growing corn for silage, deciding "it's cheaper to buy it from the farms in the Valley" where it's easier to grow. But the economics of the Fellows' operation still didn't make sense. Like so many in their situation, they almost went under.

Around 1990, Mark said, he noticed how "in summer, when the cows were on grass, the checkbook was overflowing. And in winter it was harder." After trying his brainstorm out on his vet (the vet playing devil's advocate), "I decided to go seasonal."

Mark called that decision "the best thing I ever did," and said it had some interesting effects. "It made me like a big farmer. I have a big group of cows freshening at the same time, a big group of calves I'm raising of the same age. The cows come into heat about the same time, and, instead of watching them all year-round, in May and June I take my granola into the field and watch them for signs of heat as I eat breakfast." In the late fall they're dried up all at once, and "Then I get four months off." (Maintaining a dry herd over the winter basically involves hauling hay and cleaning the barn—easier than the summer regimen of milking, manure-spreading, breeding, fence moving, haying).

After seasonal milking had settled into a pattern, "the next step was to become organic," Mark said. "We were profitable selling commodity milk. It's expensive to make the transition. We made the change because we could."

---

## Habitat

Housing for organic dairy animals may be a barn with tie-stalls, a free-stall setup, bedded or deep litter pack, or something as simple as a shade/windbreak shelter. Much of this depends on the local climate, preexisting infrastructure on the farm, type of livestock, and the seasonal nature of production and livestock needs.

Most dairy farms converting to organic keep their existing housing system. This is often an old-style tie-stall barn or loose housing with a milking parlor. Modifications to existing housing to prevent health problems become more of a priority on an organic farm, where reliance on pharmaceuticals is not an option. Improvements to housing can reduce unnecessary stress, poor ventilation, or environmental illnesses. When capital and opportunity allow, some organic dairy farmers build loose housing, often on bedded packs or deep litter. During the grazing season, most organic dairy farms make minimal or no use of the housing, other than for milking times. Before making changes to existing housing, determine what the needs are:

- What times of the year is housing needed?
- Is the bedding dry?
- Is there enough bedding?
- Is there easy access to drinking water?
- Are there manure runoff issues, and is roof-water runoff controlled well?
- Is there adequate ventilation to avoid respiratory problems?
- Are the animals lying down and chewing their cuds, or do they appear stressed?
- Is there adequate shelter from wind, rain, or other adverse conditions?
- Is the housing large enough for the number of animals?
- Is there enough room for them to exercise and move around comfortably, and do they have access to a barnyard or outdoor turnout area in the non-grazing season? Note that daily outdoor access is required for all ruminants over six months of age during the non-grazing season.
- Are the lanes and barnyard well drained, or excessively muddy?
- Does the area allow animals to engage in healthy social relationships with each other and for submissive animals to get away from dominant animals?
- Is there enough space for all animals, including submissive ones, to get enough feed?

## Young Stock

There are many ways calves are reared on organic dairy farms. However, some basic principles apply to all rearing systems for young stock:

- The health of the calves starts with the health of the dam. A healthy mother, bred to a well-selected bull, fed a balanced diet in a non-stressful environment, will produce the healthiest possible calf.
- As soon as the calf is born, make sure it gets enough colostrum.
- Raising growing calves on quality whole milk and quality forages will reduce calf mortality. Note that feeding whole organic milk is required in the organic standards.
- Prevent calf scours or excessive parasite loads.
- Design well-ventilated housing or adequate shelter to reduce risk of pneumonia.

Most dairy farms feed whole milk, hay, and grain to calves, and wean no earlier than eight weeks.

Some farms raise calves on milk for much longer than eight weeks. An increasing number of farmers feel that a calf fed whole milk for at least three months, then gradually weaned, will be better grown and able to withstand the stress of weaning and parasites. Milk can be fed from buckets, bottles, or sucker buckets with many nursing teats for groups, and some farms raise calves on nurse cows or leave them with their mothers for a period of time. Housing can include pens, hutches, tie-stalls, or individual pens. Some put calves on pasture at birth, and some wait until they are six months old. If the existing system is working, keep it. However, if there are problems, switching to a different system is worth investigating. During the non-grazing season, daily outdoor access is required for all animals over six months, and during the grazing season, pasture is required.

Preventing scours in calves is essential, as this is one of the main causes of death in young calves. Prevention measures include good nutrition, adequate colostrum early in life, clean housing, no overcrowding, making sure dams are healthy, the use of homeopathic nosodes or vaccination, probiotics, herbs, and other preventive treatments.

**Young stock. The health of the calves starts with a healthy mother cow bred to a well-selected bull. In raising a calf, the key is prevention: feeding a nutritious diet in a stress-free environment and observing the animal closely for its changing needs.**

Transitioning farms that have been feeding a calf grain with a coccidiostat or have been regularly using antibiotics and parasiticides may be in for some difficulties. If a coccidiostat or antibiotic is used to save the life of a calf, the animal must be sold as nonorganic and cannot remain in the organic milking herd. For those farms that have been relying on antibiotics and medicated feeds, it is essential to find a new calf-rearing system that prevents infection with coccidiosis or other organisms. Refining management techniques and giving supplements to improve immune system function will be important.

## Herd Health

Good nutrition can prevent disease . . .
most of the time, but not always.
—Richard J. Holliday, DVM, *Fundamentals of Holistic Animal Health*

Prevention and early intervention on an organic dairy farm is essential for herd-health management. Such attention will save money and time spent treating sick cows. In general, a farm will experience the fewest animal health problems if it has good soil health, high forage quality, offers a balanced ration with adequate DMI, provides low-stress housing and handling, and follows good sanitation and milking procedures. Incorrectly balanced rations, overfeeding, underfeeding, or giving poor quality feeds set the stage for health problems. The stress of poor ventilation, lack of a dry place to lie down, overcrowding, poor milking procedure, stray voltage, or stressful handling practices can also add to problems. Most cows can tolerate a few challenges to their health. But if they are stressed for a long period of time, or by many issues at the same time, their overall vitality will decline, and early symptoms of health problems will appear.

A veterinarian can plan an important role in diagnosis, as well as in developing a prevention and treatment plan using the materials and methods allowed by the organic standards. In addition to knowing which medications are allowed by the organic standards, it is also important to know which medications are allowed under the Pasteurized Milk Ordinance (PMO), and how medications on the farm need to be labeled. Some medications will require a veterinarian's label to avoid losing points on farm milk inspections.

### Observation

Over time, it is rather easy for a person sufficiently sensitive
to immerse their entire consciousness into that of the herd and
the individual animals that create it. Being with the cows
daily can become a long-term meditation.
—Hubert Karreman, DVM, *Treating Dairy Cows Naturally*

Keen observation and early intervention are particularly important when relying on alternative treatments. This is not because there are fewer options available to treat livestock on an organic farm, just that the treatments are most effective when used early.

A healthy cow has a heart rate of 70–80 beats per minute, a rectal temperature between 100.5°F and 102.5°F, and she will breathe about 30 times a minute. She should have good rumen activity, visible from rumen fill and movement as well as cud chewing. A great deal of information about the health of the animal can also be assessed by monitoring changes in manure texture and aroma. Observation of eyes and mucous membranes can show if there is dehydration, fever, or other problems. Behavior to watch includes how a cow stands, walks, and interacts with others, how she competes for feed, if she is off-feed, and which plants or feeds she chooses to eat or not eat. In addition to objective observations, don't overlook the intuitive skills that develop in the close relationship between farmer and cow.

## Prevention Checklists

### *Calf Diarrhea*

- Early and adequate, high-quality colostrum
- Clean housing and avoiding overcrowding
- Preventive administration of probiotics or vaccines
- Good quality nutrition
- Dry bedding

### *Pneumonia*

- Correctly ventilated housing
- Good nutrition and adequate DMI
- Dry bedding

### *Mastitis*

- Good milking procedure
- Good nutrition with particular attention to minerals
- Probiotics, vaccines, and organically approved serum products (such as a whey product)
- Good sanitation; clean, dry bedding

- Testing individual cows for SCC (somatic-cell count) and pathogens and consider using DHIA testing (Dairy Herd Improvement Association)
- Dry-cow care and good drying-off procedure
- Elimination of any stray voltage problems in the milking area or barn

*Milk Fever*

- Mineral balance of soils
- Mineral content of forages and rations (pay particular attention to potassium level of the forages)
- Body condition, good but not over-conditioned

*Foot Problems*

- Good surfaces in lanes and barnyards
- Avoid overfeeding concentrates and protein
- Reduce rumen acidosis
- Balanced feed, particularly minerals
- Nosode use if there is a known problem
- Foot baths and foot trimming if needed

*Reproductive Problems*

- Good nutrition, including a balanced mineral program
- Avoid over-conditioned or under-conditioned cows, particularly in the months prior to and after breeding

*Internal Parasites*

- Graze clean pastures with most susceptible animals (young stock)
- Include some high-tannin plants in pastures or hedgerows if possible (bird's-foot trefoil, chicory, etc.)
- Maintain good soil fertility
- Good nutrition, adequate DMI, and a well-balanced mineral in ration
- Consider feeding milk for at least three months and wean slowly to be sure calves have good body condition

*External Parasites*

- Fly traps, sticky tapes in barn
- Good sanitation in and around barn and manure-storage areas
- Encourage beneficial wildlife with birdhouses and habitat
- Release predator wasps

## Milk Quality

Drying off is a critical time for udder health, and any extra care given at this time will pay big dividends throughout the next lactation. Prepare the cow for the stressful transition from lactating to non-lactating by using your favorite herbs, homeopathic preparations, colostrum products, acupuncture, or others to boost her immune system and help relieve stress.
—Richard J. Holliday, DVM, *Fundamentals of Holistic Animal Health*

As somatic-cell count (SCC) increases, milk quality decreases. This lower quality results in lower cheese yield, shorter shelf life, and a lower pay price for fluid milk. Maintaining an SCC below the quality-premium level of your milk buyer not only increases income, it also means cows are healthier. SCC varies with age, stage of lactation, and health of the cow. A high number may indicate mastitis, poor immune function, udder injury, stress, or other factors. Some ways to lower SCC include:

- Good and consistent milking procedure
- Pre- and post-dipping (check to be sure that the materials are allowed by your certifier)
- Stripping each teat and visually checking each quarter for clots or inflammation
- Use of DHIA or a mastitis test to monitor changes in SCC
- Keeping the milk of high-count cows out of the tank
- Use of a salve or other skin conditioner to keep the udder and teat skin healthy
- Good ration balancing and the use of supplements and management practices to improve overall cow vitality

A consistent milking procedure that ensures a good letdown, well-maintained milking equipment, low-stress handling of animals at milking time, and good sanitation are important in milk quality and prevention of udder health problems. An effective, consistent method of drying cows off and caring for them during the dry period and at calving will favorably affect SCC and udder health.

The best method to dry a cow off is to stop milking the cow abruptly, rather than the older method of decreasing the interval of milking to once a day, then every other day and so on. The cow needs to have a tight udder so that her system gets the message to stop making milk. After milking is stopped, wait four or five days (until the size of the udder begins to decrease naturally as milk production ceases), then check each quarter for signs of infection. Treat with an allowed product if needed. If you strip the quarters out at this time, use a teat dip and good sanitation. Continue to monitor the udder during the remaining dry period, stripping the quarters out only if there appears to be a problem. For cows that continue to drip milk, it may be useful to continue using a teat dip, both at drying off and prior to calving.

Pay close attention to the cow as she begins to bag up prior to calving. It is much easier to treat the cow for any potential udder infection before calving than to wait till her udder is full of milk. Some farmers also pre-milk certain cows, particularly if they have a tendency to be very high producers, are leaking milk, or get edema. With pre-milking, colostrum will still be produced shortly before calving, and can (and should) be saved for the calf.

## Internal Parasite Management

> An animal without worms is not an ideal to strive for at any cost, at least not in organic farming. An animal that never has worms cannot develop resistance, and is thus extremely vulnerable when exposed to a parasite.
>
> —Jean Duval, *The Control of Internal Parasites in Cattle and Sheep*

Parasite management remains one of the biggest challenges to organic livestock production. Most "natural" treatments available do not work the same way as chemical dewormers, so prevention is essential. All farms and all livestock have parasites, so the challenge is to manage the animals in such a way that the parasites do not create health problems. It is useful to learn the parasites' life cycle: how they infect livestock and what the parasites' most vulnerable life-cycle stage is. There are many different types of

Nematode life cycle. Adult worms live in the digestive tract of livestock. An egg passes in manure onto the pasture; a young worm develops there in the egg; it hatches as a larva; the larva develops to the infective stage and is swallowed by a grazing animal, and the cycle repeats.

> Some of the natural remedies being used on organic dairy farms include pumpkin seed, black walnut, wormwood and other Artemisia species, male fern, garlic, cloves, plants with high tannin content such as woody species and brambles, and many other plant species.

internal parasites, each with different host organisms and life cycle. Fecal testing can determine which species of parasites are present in the herd. It is also interesting to consider the role of parasites, both on a farm and in wild animal populations.

Resistance to or tolerance of parasites varies with breed, age, previous exposure, selection, immune function, soil quality, and feed quality. Young stock, with immature immune systems, are more susceptible to parasites than mature animals. Among ruminants, goats are the most susceptible, followed by sheep, and cows are the least susceptible. Grazing management to provide clean (noninfective) pastures to the most susceptible livestock, as well as good sanitation, good nutrition, providing ample colostrum soon after birth, and selection of hardy animals are all basic in prevention.

To treat internal parasites with herbs or other natural remedies will require more than just switching from pharmaceutical dewormers to a natural dewormer. The natural products work differently and require a more holistic approach with a focus on prevention of infection and creating a healthy animal with a highly developed immune system. Once infection loads are high, don't count on "natural" treatments alone. There is a review of literature on parasite treatments in Karreman's *Treating Dairy Cows Naturally*. There is also information in the health chapter of *The Organic Dairy Handbook* from NOFA–NY.

Fecal egg counts can be a useful tool in parasite management, but don't rely on them to tell everything about the parasite loads livestock have. Not all parasite species produce many eggs, and there are seasonal variations in egg production. If you do decide to use fecal egg counts, use them at critical times of the year, the same time each year, and consider them just part of the overall management plan.

## Parasite Prevention

Here are some guidelines for preventing parasites in your herd:

- Pay close attention to vitamins and minerals in the ration.
- Keep livestock in good condition, and don't allow them to become too thin.
- Minimize infection by using feeders that animals can't get manure into, and avoid ground feeding.
- Minimize or eliminate infection from pasture by keeping the most susceptible animals out of infected pastures.

Parasites will live, and continue to be infective, for about a year (longer in warm, moist climates) after they are shed in manure on the pasture. Animals that shed the most eggs are usually mothers after birthing, or young stock. Sheep and goats have (mostly) different parasites than cows, so multispecies grazing can be part of the strategy to create clean pastures. Other ways to decrease parasite load in a pasture include harvesting hay, fallowing an area, and rotating with an annual crop.

Here are additional recommendations for preventing parasites:

- Including some high-tannin plants (e.g., bird's-foot trefoil, forage chicory, brambles, many woody species) in hedgerows and pastures seems to reduce parasite loads.
- Farms with a lower stocking rate generally have lower parasite loads.
- Not allowing animals to graze the pasture down shorter than 2 inches, particularly in wet weather and on more heavily infected pastures, can reduce infection.
- Improving soil fertility and the use of compost seem to help decrease problems.

## Flies

Fly management begins with an evaluation of the problem. Is the fly problem merely annoying, or is it affecting livestock production and welfare? Are flies in the milking area creating a hazard for the people doing the milking (kicking cows and swinging tails)? As with internal parasites, prevention is the easiest place to address fly problems. Some preventive methods include these:

- Elimination of breeding areas near manure storage and barnyards
- Release of fly-predator insects (can be purchased by mail)
- Sticky traps and other types of fly traps in barns and milking areas
- Use of birdhouses and other ways to encourage fly-eating wildlife

Some fly sprays are now approved for use on organic dairy farms. Before use, be sure to check with your certifier that all the ingredients (both active and inert) are allowed. Some farmers make their own fly sprays, using ingredients such as vegetable oil and essential oils.

## Treatment Products and Methods

> The success of the holistic approach requires a change in perspective and the development of a holistic outlook towards livestock management and disease control. It is not as simple as merely substituting a "natural" alternative therapy for a "toxic" drug. The principles behind the success of holistic therapy go much deeper than the characteristics or source of the medication.
> —Richard J. Holliday, DVM, *Fundamentals of Holistic Animal Health*

Many dairy farmers find that, after the transition and the first few years of organic management, they face fewer herd-health problems. This is likely because management practices are taking effect: culling and selection decisions, improved nutrition, lower stress, and better prevention, observation, and early intervention. However, no matter how much care goes into prevention, all organic dairy farmers will still have to deal with some herd health-care issues.

There are now several books available on organic health-care treatment options. This handbook will give a brief overview of the most commonly used methods. You'll want to do further reading and research specific treatments and methods beyond what is presented here.

It is important to know what health-care treatments are allowed, which ones are prohibited, and which ones are allowed with certain restrictions. Your certifier will require that you keep records of all the health-care treatments and materials used. You must also have your certifier review all the products *before* you use them. This includes conventional treatments and all the alternative treatments.

The national organic standards allow most natural (non-synthetic) health-care products unless they are specifically prohibited. The standards allow synthetic health-care products *only* if they are on the national list. Because some synthetics are allowed, a few treatments on an organic farm will be the same as on a nonorganic farm. Dehydration, for example, will most likely be treated with an IV or drench with an allowed material, and some of the regular first-aid treatments will be the same. The differences will appear most often at the next level of treatment. When a hormone, antibiotic, or other prohibited synthetic product would be the choice on the nonorganic farm, the organic farmer needs to be familiar with the natural alternatives that can be used. Also note that the organic standards require that when an alternative treatment fails and animal welfare is in jeopardy, you must provide necessary medical treatment, even if that animal must then be removed from the organic herd and sold as nonorganic.

During the past several years, the types and numbers of organic treatment options and products have increased. These treatments include synthetics, which have been added to the national list, as well as alternative treatments. Although more information is now available on the effectiveness of some of these alternative products and how they are best used, there is still a great deal of research needed on this subject.

When buying any health-care product, look carefully at the labels and ingredients to be sure it is allowed by your certifier.

- Look at both the active and inert ingredients to make sure they are allowed.
- Check to see if all ingredients are listed on the label.
- See if the product is listed by the Organic Materials Review Institute (OMRI).
- Ask your certifier if the product is allowed.

Here is a partial list of some of the health-care products and ingredients that are currently allowed for use. Note that some of these come with specific restrictions concerning when and how you can use them, as well as milk withholdings that may be longer than the usual amount of time. These products include the following: flunixin (Banamine); furosemide; aspirin; lidocaine and procaine; xylazine; tolazoline; vaccines; magnesium hydroxide and magnesium sulfate; oxytocin; ivermectin; poloxalene; atropine; butorphanol; electrolytes; and alcohols, chlorine, and iodine for disinfecting.

### Herbs

The use of medicinal herbs or botanicals to treat illness is nothing new. People have used plants as medicine since before recorded history, and this form of treatment is now being used widely and studied more, particularly as more strains of antibiotic-resistant bacteria and other pathogens emerge.

Stephen Harrod Buhner, in his book *Herbal Antibiotics*, says, "research has revealed that instead of being a quaint quackery of our forefathers, many herbs possess strong antibacterial qualities, in many instances being equal to or even surpassing the power of antibiotics"(p. 19). Herbs, with their complex mix of many constituents, are not subject to resistance from bacteria the same way penicillin or other antibiotics are.

There are plants that act as diuretics, relieve pain, improve immune function, reduce inflammation, and more. Pastures and hedgerows contain an array of plants that provide many types of both nutrition and medicine. Livestock may "self-medicate" while grazing with plants like dandelion, plantain, burdock, goldenrod, hawthorn, nettle, peppermint, shepherd's purse, yarrow, and more.

These plants can also be harvested and used to treat livestock. Most dairy farmers don't have the time to grow and process herbs, but there

Some medicinal herbs used on organic dairy farms include aloe, burdock, coltsfoot, comfrey, dandelion, echinacea, elecampane, garlic, nettle, plantain, raspberry, St. John's wort, wormwood, yarrow, and many others.

are companies that make and sell medicinal herb products for livestock. Some medicinal plants are nontoxic even in large amounts, and are even grown as forage species (dandelion, chicory, red clover). These plants can be grazed by livestock, fed dried or fresh in the barn, or can be used in tincture form or as a tea, or infusion in a drench. Some of the more powerful medicinal plants can be toxic; however, some may be used as medicine in the correct dose. Many of these more toxic plants are used in their extremely dilute homeopathic form (aconite, phytolacca). Using plants like these is one more reason to have a good working relationship with a knowledgeable veterinarian who can help develop a safe and effective treatment plan.

Medicinal herbs can be given to livestock as:

- Fresh or dry plants fed directly or grazed—*note that any feed fed to organic livestock needs to be certified organic*
- Dry plant material used to make a strong tea (infusion or decoction) that is then used as a drench
- A salve or poultice (check the ingredients in purchased salves to be sure they are all allowed)
- Tinctures, made by adding the plant (dry or fresh) to alcohol or vinegar, allowing it to sit for a period of time (generally at least six weeks), and then straining it off; often used as a drench

Talk with your veterinarian, or refer to the resource section at the end of this book for more information on this topic.

## Homeopathy and Nosodes

Homeopathy has been in use since the mid-nineteenth century. The word, which is derived from the Latin word for "similar," is based on the idea that "like cures like," and was developed by Dr. Samuel Hahnemann. Homeopathic remedies are unlike any of the other health-care products because they are very dilute, and the more dilute remedies are actually more "potent." Homeopathic remedies work by strengthening the vital force of the animal rather than treating a pathogenic organism or symptom.

There are two general ways that organic farmers use homeopathy. One is to use a simple "cookbook" approach, where for ailment "A," you administer a particular remedy. This is the practical way in which many farmers apply the rather complex homeopathic form of medicine. Although very different from the approach designed by Hahnemann, it appears to work on some farms. The more classic approach to homeopathy is a detailed collection of information about symptoms and the individual that is precisely matched to the remedy. This approach is used by some well-trained veterinarians, as well as by farmers who have taken classes in classical homeopathy, or are participating in local homeopathy study groups. Over time, a more in-depth understanding of homeopathy will allow a more effective use of the remedies.

For a farmer getting started with homeopathy, there are an increasing number of veterinarians trained in this method. There are also several good novice-level books that come with a small "kit" of common remedies. The homeopathy kit should be stored away from excess heat, electrical fields, and aromatic medicines (like liniment rubs).

In addition to regular homeopathic remedies, there are also homeochords, which are blends of several potencies, and are another good choice for the beginner homeopath. There are also homeopathic nosodes available from veterinarians to help in management of particular problems including mastitis organisms, foot rot, and scours. A nosode is a homeopathic remedy generally made by a veterinarian for a specific pathogen or problem.

To administer a homeopathic remedy, ensure that it contacts a mucous membrane. A liquid spray can be misted on the nose, pills can go into the animal's mouth or in the vulva. When treating a whole group, the remedy can be added to their water. The more rapid the onset of the ailment, the more frequent the doses.

Once you become more familiar with homeopathy, it may be useful to understand the constitutional homeopathic remedies. These can be

> Homeopathic remedies commonly used on organic dairy farms include Aconite, Arnica, Carbo veg, Hepar sulf, Hypericum, Lachysis, Nux vomica, Pulsatilla, Phytolacca, Sepia, Silica, Sulfur, and others.

Treatment products and methods—ancient and continuing. A few of the herbs used on organic farms are dandelion, echinacea, plantain, and St. John's wort, all pictured here. People have been utilizing the medicinal qualities of plants since before recorded history. Scientific studies have shown that some herbs possess strong anti-bacterial qualities.

particularly useful in treating a chronic illness. The aim is to match a remedy to an animal's personality, behavior patterns, responses to its environment, common ailments, or symptom. Karreman's *Treating Dairy Cows Naturally* has a good description of the constitutional approach to homeopathic remedies. *Homeopathy for the Herd*, by Dr. Edgar Sheaffer, is a good book for farmers who want to start using homeopathy on their herd.

### Probiotics

Probiotics are beneficial live microorganisms that are fed preventatively or as part of a treatment. A healthy digestive system contains many different types of organisms that are a necessary part of the immune and digestive systems. A healthy animal eating a high-forage diet including pasture should be able to get and maintain the right population of these organisms, but a stressed or sick animal and calves can benefit from supplemental probiotics. Some organic dairy farms use probiotics preventatively by giving them to all newborn calves. This can be particularly helpful if they are at risk of scours. Probiotics are available commercially and are sold under a wide range of names and brands. Many are mixed with other ingredients including minerals and vitamins. Take care that the mixture doesn't contain any prohibited additives.

### Vaccines

Vaccinations are allowed for use on organic livestock. The decision about which vaccinations to use, how many, or whether to use them at all should be made based on the individual farm and the health challenges that the herd has. On a farm with a closed herd, good biosecurity, and no history of illnesses, some veterinarians suggest that adding a vaccination protocol is unnecessary. Dr. Richard Holliday writes, "Vaccinations may increase resistance against a specific organism but [do] little to elevate the animal's vitality to the health enhancement level."

For farms that have a vaccination program that is working well, there is little incentive to change. Farms struggling with disease outbreaks may want to consider adding vaccines to the overall preventive care of a herd. For farmers who don't wish to use vaccines, good preventive practices, including biosecurity and the use of homeopathic nosodes, may be an alternative.

## Other Treatments

There are several products that are blends of herbs, vitamins, and minerals. There are also immune-stimulant products such as colostral-whey, hyperimmunized serum, or brand-name products such as Immunoboost. There are several companies now making products like these, and some now have been approved for use by certifiers or OMRI. Check the ingredients before using any of them. There are many "natural" health-care products sold that contain prohibited ingredients. Remember to ask your certifier about the product before you use it.

Cider vinegar, an old treatment for many ailments, is receiving some deserved attention again. It was first made popular by Dr. Jarvis in his *Folk Medicine*, where he recommends feeding two ounces a day as a treatment or preventative.

Acupuncture and acupressure can be used both for treatment and for diagnostics. Acupressure works best with dairy animals that can be handled easily. The technique is best learned from someone else, but acupressure can also be learned and practiced by studying the charts and practicing on your own animals. Most of the commonly used points for a cow are located along the spine, and you can find copies of charts in Karreman's *Treating Dairy Cows Naturally* and Holliday's *Fundamentals of Holistic Animal Health*. Acupuncture, developed in China, is based on the relationships between certain organs and "points" on the body surface. These associated points can be used both to indicate which organs are not functioning well, and to stimulate their healing.

## Specific Treatment Suggestions

The following list of problems and treatments is adapted from "Holistic Animal Husbandry" by Jerry Brunetti, director of Agri-Dynamics (www.agri-dynamics.com). See the NODPA Web site (www.NODPA.com) for the full-length version of this article.

Not all of these products may be allowed for use by your local organic certifier. Always check with your certifier before using a new product. In addition, it's important to have a good working relationship with a veterinarian for diagnosing and treating these issues.

### *Mastitis and Somatic Cell Count*

- Trace mineral analysis: zinc, iodine, selenium, copper
- Approved hyperimmunized serum or colostral-whey products
- Antioxidant vitamins (C, E, A)
- Mash/gruel of vinegar, molasses, bran, beet pulp
- Herbal extracts: cayenne, ginger, mint, licorice, echinacea, garlic
- Udder massage—with an approved liniment or salve

### *Ketosis*

- IV treatment with an approved energy source
- Drenches with liquid energy sources; note that propylene glycol is prohibited as an active ingredient
- B-vitamins, especially niacin (12 grams/day)
- Crude liver extract
- Probiotics to help appetite
- Herbal extracts to help digestion (chicory, ginger, fenugreek, cayenne, licorice, peppermint, fennel), herbs for liver health (dandelion leaf, burdock root, yellow dock root, milk thistle seed)

### *Milk Fever*

- IV electrolytes with calcium/dextrose (also use vitamin C to assist calcium transport)
- Drenches with calcium (note that calcium propionate is prohibited); vitamin D; phosphorous (calcium phosphate); zinc (sulfate, chelate); cobalt (sulfate chelate); vitamin B-12

### *Udder Edema*

- Vitamin injections: B-6, B-12
- Diuretic herbs: dandelion leaf/root, celery seed, juniper berries, gravel root, cleavers (note that a veterinarian may also be able provide an allowed conventional diuretic if needed)
- Udder massage with "hot" liniment (camphor, peppermint, capsicum, etc.)
- 1 pint of strong coffee

### *Calf Scours*

- Demulcent herbs: psyllium, comfrey, slippery elm, mallow; astringent herbs like raspberry leaf, blackberry root, bayberry; antimicrobial herbs like Oregon grape, bearberry, peppermint, eucalyptus, garlic, thyme, etc.
- Activated carbon, clays (montmorillonite, bentonite, attapulgatite, etc.)
- Electrolytes with buffers
- Probiotics/enzymes
- Approved hyperimmunized serum or colostral-whey products

### *Foot Rot*

- Minerals/fiber to correct acidosis
- Increases in iodine, zinc, and sulfur
- Topical foot bath with copper and zinc sulfate, and iodine (note that a veterinarian may also provide an allowed salve, or you can make one using allowed ingredients such as iodine, Epsom salts, sugar)
- Good pasture—grass factors and nutritional/medicinal herbs
- 4–6 ounces montmorillonite clay per head per day (buffer silica, calcium, magnesium, trace elements) in feed

### *Reproduction Problems*

- Correct acid/alkaline balance
- Test for MUN/BUN (milk urea nitrogen and blood urea nitrogen); no excess protein in diet
- Adequate A, E, copper, cobalt, zinc, iodine, B-complex, selenium, iron
- "Grass factors"—enzymes, carotenoids, vitamin E, trace minerals, nutritional and medicinal herbs
- Assessment of cow's body condition

### *Respiratory Problems*

- Vitamin A, E, C, B-complex injections
- Approved hyperimmunized serum or colostral-whey products

- Herbs: hyssop, eucalyptus, peppermint, thyme, fennel, fenugreek, garlic, elecampane
- Removal of animals to draft-free, well-bedded outside environment.
- Laxatives, supplement diet with bran, beet pulp, molasses mash; grass/hay
- Note that a veterinarian may also be able to provide other helpful treatments such as Banamine.

### *External Parasites*

- Good ventilation, direct sunlight
- Essential oils in an approved mixture; apply by rubbing and spraying

### *Internal Parasites*

- Prevent infection by grazing most susceptible young stock on clean pastures where there aren't parasites
- Feed adequate levels of minerals
- Herbs: garlic, clove, wormwood, wormseed, black walnut hulls and leaves, neem drenches of these plants for individual treatment; (note that Ivermectin is allowed for use; however, it comes with use restrictions, and cannot be used routinely)
- 4–8 ounces of apple cider vinegar diluted in 20 gallons of water on a daily basis
- Access to plant diversity, especially woody hedgerows and conifers

## FEATURED FARM:
## Hawthorne Valley Farm, Ghent, New York

Hawthorne Valley Farm is certified organic and certified biodynamic by Demeter. The dairy herd is made up of about 65 Brown Swiss and Swiss crosses. The barn is a tie-stall, with a pipeline milking system. The farmland is rolling hills in the

Taconic range of New York and includes pasture, hayfields, and land used for production of vegetables and cover crops that are sometimes grazed.

During the summer the milking herd is intensively grazed, getting fresh pasture after each milking. Pastures are sometimes clipped in the spring, and as pasture growth slows down, some hayfields are grazed. In winter, dry hay is fed to the cows instead of silage owing to milk flavor for the raw-milk customers and the cheese-making business. Fertility of the pastures and hayfields is accomplished through good grazing management, use of legumes, occasional use of lime, and the spreading of composted manure.

The cows are fed a small amount of grain (well under 5 pounds per day) in the barn, supplemented with minerals including kelp and Redmond's mineral salt. Whole grains (mostly corn) are bought and ground before feeding. The hope is that in the future there will be more locally grown corn or other grains available.

The farm raises all its own replacements, but does buy a bull every few years. Young stock are started by leaving the calves with the mother for about three days. They are then moved to a pen where they are fed whole milk and hay. Young stock go out to pasture at about five or six months, when they are well grown and have developed some resistance to parasites. They are grouped by age, and at 18 to 19 months are put in with the bull. Some heifers are crossed with an Angus bull to supply beef animals.

Herd health treatments when needed are mostly homeopathic remedies, or a bottle of calcium as needed for milk fever. Fewer milk fevers have been observed since closer attention has been paid to the nutrition of the dry cows. Calf diarrhea remains one of the larger challenges, along with mastitis. Culling decisions are often based on SCC or staph aureus test results.

Hawthorne Valley has a license to sell raw milk, and the remainder of their milk is processed on the farm into yogurt, fresh cheeses, and aged cheeses. New York is one of the states that has licensed raw-milk dairy farms. The fresh cheeses (quark and bianca) and yogurt are pasteurized, and the aged cheeses are raw-milk cheeses (aged more than sixty days to meet regulations) and include Edam, alpine cheese, and cheddars.

In addition to the dairy farm and dairy processing, Hawthorne Valley also produces beef and raises pigs (fed whey). The farm also includes a large vegetable operation, on-farm educational activities, a bakery, and a farm store. The educational activities include a visiting student program that allows students to spend a week on the farm doing chores with the farmers, staying in dormitory housing, and eating food cooked and raised on-farm. There is also a two- to four-week summer camp for nine- to fifteen-year olds. The farm and all the other activities are part of the Hawthorne Valley Association, a not-for-profit organization "committed to fostering new insights and capacities in agriculture, education, and the arts."

Markets for Hawthorne Valley cheese, meats, baked goods, and vegetables include a CSA (community supported agriculture), the Hawthorne Valley Store, and green markets in New York City. Some yogurt is sold through a distributor to other parts of the country. Demand for their dairy products is so high that they are now buying milk from another organic dairy farm, and would like another biodynamic dairy farm to start up in their area to provide them additional biodynamic milk.

The farm brochure reads, "Ideally, the biodynamic farm is a self-sufficient and self-contained ecosystem that can fulfill its needs from within," and this farm is indeed closer to being a self-sufficient ecosystem than most farms are. The mixture of livestock producing value-added products, whey for pigs, and manure—which is composted and used on the vegetables—creates a challenging, labor-intensive farm that seems to be evolving successfully to meet the ever-growing demand for its products and services while also taking care of the people who work there.

---

five

# Marketing

## Selling Fluid Milk

Unless you plan to develop your own products and market them, it is critical to find a buyer for your milk before you start producing it. Local access to a market for organic milk varies widely. The milk pay-price offered by buyers in different areas varies also. Some questions to consider include:

- Is there a local buyer?
- Are they interested in buying your milk (based on location of your farm, milk quality, quantity)?
- What is the organic milk price, with what adjustments for quality or components?
- Is it a cooperative or a proprietary business, and what are the contract details?
- What are the hauling charges or other costs?
- What market does the milk go to, what is their long-term marketing plan, and how stable does the demand and pay price seem to be?
- Are they willing to sign a contract or letter of commitment to buy your milk before you incur the cost of transition to organic?

## Milking Systems and Equipment for Fluid-Milk Production

Most organic dairy farmers convert an existing dairy farm to organic, and are familiar with state regulations and licensing requirements to produce and sell milk. For a new farm, contact your state's department of agriculture for requirements for facilities (before you build), inspections, and licensing.

Here are some questions to ask yourself as you design a new milking system or modify an existing one:

- Will it comply with all the necessary regulations?
- Are the cows presently calm as they come in, get milked, and move out?
- Are there any existing or potential stray-voltage problems?
- How long does it take to milk, and how could milking efficiency be improved?
- Are you feeding grain or minerals in the parlor?
- Do the facility and equipment allow for a good milking procedure and preventive care of animals?
- Are there any milk-quality problems caused by the existing system?

## Producing and Marketing Value-Added Dairy Products

> Direct marketing is best; start slowly and see if it works. See if you can make a go of it before you turn all your milk into cheese, or buy a year's supply of yogurt cups and find that the market is already flooded. . . .
> —Jeannette Fellows, Chase Hill Dairy Farm

Consumers in some areas are willing to pay a premium for high-quality, organic, farmstead, grass-fed products. Making and marketing cheese, butter, bottled raw or pasteurized milk, yogurt, cream, or other products is an alternative to selling only fluid milk. It can be a way to add value, assure a steady income, and allow an organic dairy farm to sustain itself where there may not be a buyer for organic bulk fluid milk.

However, the development of products, markets, business plan, facilities, and acquiring the equipment needed for both production and marketing takes money, skills, and time. Before making this major investment, spend time visiting other dairies that do on-farm processing, attend some workshops, and revisit your farm-family goal.

- Is this new enterprise consistent with your overall farm and family goal?
- What are your marketing goals?
- Is there an adequate market in your area for the product?
- Is there sufficient skilled labor available?
- What equipment and facilities are needed, and how will you pay for them?
- What type and scale of equipment is needed for the different products?
- What is the shelf life of the product?
- What are the regulations and the licensing and inspection requirements?
- Are you able to produce the high-quality milk needed for production?
- Do you have the capital and labor resources available to start up the enterprise?
- Do you have access to people and information to help develop the products and market?

### Regulations

Regulations vary from state to state, and are different for different types of products. Contact your state department of agriculture (or health department in some states) to find out what the local requirements are. For the production of organic dairy products, organic certification of both the farm and the processing plant will be required, so contact your local certifier for the additional standards information.

### Raw-Milk Cheeses, Pasteurized Cheeses, and Other Products

For cheese sales, pasteurization (according to the pasteurized milk ordinance, or PMO) is generally required for all cheeses aged less than sixty days. These types of cheeses include most brie, camembert, fresh farmer's cheeses, and other fresh cheeses. Pasteurization is also required for other non-aged products including yogurt, butter, and cream.

Cheeses aged more than sixty days can be made from raw milk in most areas. These aged cheeses that may be made from raw milk include cheddars, Gouda, Swiss, and many others. There are variations from state to state in how the PMO has been adopted, so research for each type of product is necessary.

Each of these dairy products present different production challenges and will require different types of facilities and equipment for production and storage. The storage needs and shelf life must be seriously considered in designing the marketing plan. For the production of any of these products, before you begin it is essential that you have an adequate supply of high-quality milk.

### Facility and Equipment

The size and design of your facility will depend on the type of products you decide to make. You generally need one room for raw-milk storage and for washing milking equipment. A separate room is usually required for making the products. A separate cheese cave will be needed for any aged cheeses. Room will also be needed for storage of supplies and for packing and shipping. There are specific requirements for sinks, floor drains, wall and floor materials, bathrooms, and much more. Contact your state's department of agriculture for more information.

The types and sizes of equipment for making value-added dairy products vary widely. For pasteurization, there are vat or high-temperature short-time (HTST) pasteurizers. The vat pasteurizer may also be used as a vat for making cheese. There are many types of cheese vats and kettles with different heat sources and various levels of automation. Other equipment used may include a cream separator, butter churn, vacuum packer, slicer, bottling and capping machine, and more. Lab-testing equipment and tools may be needed to test the quality of the milk or other products including fat content, moisture, SCC, bacteria, pH, and more.

Some excellent workshops and classes on value-added dairy production are held regularly through various organizations and universities. In addition to attending some, it may also be useful to visit other farms and facilities before you begin designing recipes, facilities, and business plans.

There may also be a nearby facility at which to begin making test batches. This can help you develop and assess your markets and your skills as a cheese maker.

The SARE publication called *The Small Dairy Resource Book* by Vicki Dunaway is an excellent listing of resources available for on-farm dairy processing, as is the *Cream Line* newsletter. A helpful source of information on marketing is the *Legal Guide for Direct Farm Marketing* by Neil Hamilton. Useful ATTRA publications include: "Value-Added Dairy Options," "Adding Value to Farm Products: An Overview," "Keys to Success in Value-Added Agriculture," "Direct Marketing," and "Evaluating a Rural Enterprise."

### Selling Raw Milk

Some states allow the sale of raw milk, and there is growing consumer demand for high-quality, organic, grass-fed raw milk and other raw-milk products. Regulation of raw-milk sales varies widely from state to state. In some states the sale of raw milk is illegal, in other states farms can get a license to sell raw milk.

Not all state regulators create a friendly environment for raw-milk sales. The resources section at the end of this book lists information sources for both farmers and consumers interested in this topic.

If raw-milk sales are to be considered, it is essential that the farm is able to consistently produce high-quality milk.

---

**FEATURED FARM:**
**Strafford Organic Creamery, Strafford, Vermont, Earl Ransom and Amy Huyffer**

---

Earl and Amy now process and market all their own organic milk. Prior to building the processing plant, which they started in 2001, Earl and Amy shipped organic milk to Horizon, and before that they shipped to The Organic Cow. (The following is an excerpt, reprinted here with permission, from an article by Jack Kittredge in the Summer 2003 issue of *Natural Farmer.*)

This 40-cow dairy herd is primarily Guernsey. Earl and Amy keep a bull rather than use artificial insemination, and time his visits to the cows with their need to increase production in the fall, when demand is higher. So he is busiest with the herd in November and December, in order for the girls to freshen around September.

The cows are milked twice a day, at 4:30 and 4:30 in the new milking parlor the family recently added to their old 36 tie-stall cow barn. "It's a step-up walk-through 6-unit parlor," Amy explains. You don't have to get a batch of six in at once. That's good since some cows take longer to milk out than others. They step up about 18 inches, which is nice. The udders are right there so you don't have to crouch down and you can monitor their health much easier."

The farm raises all of its own forage and about 35 percent of its grain—some of that on rented river-bottom land. The cows have about 100 acres of hayfields, 35 or 40 acres of grain, and 75 acres of pasture.

Most of the pastures are divided into small rotational paddocks. . . .

Ransom prefers dry hay to wrapped because the cows seem to milk better on it. When the weather doesn't permit, however, he wraps it.

All Strafford Organic Creamery milk is bottled to order and sold to stores. Strafford's product line includes five different milks: Creamline (an unhomogenized whole milk), homogenized whole, 2%, 1%, and skim. The milk comes in quarts, half gallons, and some in pints, all glass, and all with the Strafford Organic Creamery logo. They also sell a couple of extra products: chocolate milk, half-and-half, and sometimes whipping cream. But currently most of the dairy's cream is going into ice cream. Earl and Amy make and sell ten flavors of ice cream.

six

# Record Keeping

Regardless of their source, adequate records are the
bedrock foundation of any successful business.
—Richard J. Holliday, DVM, *Fundamentals of Holistic Animal Health*

Organic certification requires that you keep adequate records to show that your product is organic. Additional records will help you track your farm finances for tax and management purposes. State regulations of milk and value-added dairy production on farms will have additional record-keeping requirements.

Record keeping is also essential for planning and good management. Record-keeping systems should be designed to be useful to you and the others on your farm. Create a system that is not so complicated that you won't do it, but be sure it contains the information you need for management and for the organic standards. While some farms find computerized records helpful, there are many handwritten systems that are also working well.

Records for an organic dairy farm required by the organic standards include:

- Milk production and quality records
- Product sales for milk, livestock, meat, forages, and all other farm-product sales
- Health care, breeding, calving, vet care, treatments
- Crop production, including harvest amounts
- Soil-fertility amendments, including manure
- Maps of all fields, including pastures
- Purchase records and invoices for livestock, health care products, fertilizers, feed, and pretty much everything that comes onto the farm
- Feeding and/or ration change records, as well as grazing records to verify you are meeting the pasture requirements

Contact your certifier to find out what additional records are required.

If you are doing any on-farm sales or making a product to sell, there is a whole additional set of record-keeping requirements. Other useful records can include:

- DHIA or other milk-quality and quantity records
- Financial records for tax purposes as well as management
- Labor records
- A written farm-family goal

seven

# Economics of Organic Dairy Production and the Transition to Organic

A farmer doesn't have to be an economist or have a business degree to run a profitable and sustainable organic dairy farm. Each farm and farmer will find his or her own system of financial management. However, a good understanding of the overall income and expense picture is needed, as well as an understanding of the additional costs of transitioning to organic practices. Milk price is not the only determinant of the profitability of a dairy farm. There are several articles on the economics of organic dairy production on the NODPA Web site (http://www.nodpa.com/production_economics.shtml). Elizabeth Henderson and Karl North describe an excellent and holistic method of financial planning in the companion NOFA manual *Whole-Farm Planning: Ecological Imperatives, Personal Values, and Economics.*

The cost of the transition period is one of the important economic considerations for a farm deciding both when and if to convert to organic. In order to transition a herd to organic, it must be managed and fed 100 percent organic feed for a full twelve months. The transition may involve cost in purchased forages, changes in forage yields, feeding a higher amount of forages (needing more silage or hay), new types and amounts of grain and associated changes in milk production, investment in new health-care products and resources, and unforeseen culling of cows that don't fit the organic system. A helpful resource in this process is *Transitioning to Organic Dairy: Self-Assessment Workbook* from NOFA–NY.

The cost of feeding organic grain while still getting paid the significantly lower nonorganic milk price for a full year can be a financial challenge for many dairy farmers, particularly if they are transitioning during a period of low conventional milk prices. Some farms, in order to avoid the cost of the twelve-month herd transition, are instead selling their nonorganic herd and buying an organic one. Some are transitioning a group of heifers, then timing the breeding so they begin calving after they are certified to avoid feeding as much grain during the transition. All of these transition options should be researched and discussed, and a transition budget should be developed, particularly if it will be necessary to get a loan to make it through the transition process. There may be some financial assistance with transition costs through NRCS programs, and sometimes milk companies looking to encourage more farms to transition offer a per-hundredweight financial incentive.

eight

# Nutritional Qualities of Grass-Fed Ruminants and Poultry

More consumers are specifically shopping for grass-fed meats, eggs, and dairy products, and, in some cases, dairy farmers who use high-forage rations may get paid a premium for their milk. Articles, books, and Web sites advocating eating dairy and meats from grazing animals are aiding in this increase in consumer interest.

Ruminants fed mostly or entirely on pasture produce meat and milk (or more specifically the fats in the meat and milk) that contain different amounts and types of nutrients than grain-fed livestock. The nutrients that appear in higher amounts in the meat, milk, and eggs of grass-fed animals include beta-carotene, vitamins A, E, and D, omega-3 fatty acids, conjugated linoleic acid (CLA), and other nutrients. If you are interested in reading some of the research on this topic, www.eatwild.com lists a number of research papers and articles on the subject, and *Pasture Perfect* has an extensive bibliography of sources. In the last few years, the *Stockman Grass Farmer*, the *New York Times*, *Acres*, and other magazines have also carried reports.

It is important to note that the *amount* of pasture relative to the amount of grain in the diet of grazing livestock has an effect on how much of these nutrients will be present in the meat, milk, or eggs. How beef cattle, dairy cows, poultry, sheep, or goats are grazed or fed supplemental grains and stored winter forages determines the nutrient content and nutrient density of the foods produced. So, if you are making claims about "grass-fed" or nutrient-content claims about your farm products, be ready to back them up with some good information. Additionally, much of the research doesn't clearly "prove" any health effect of increasing these nutrients in our diet, so be clear about that in your farm brochures or product labels.

In addition to the information about the nutritional content of grass-fed products, consumers are also attracted to some of the other benefits of grass farming, including improved animal welfare and the many environmental benefits.

The way this information will influence farm management and product marketing will vary from farm to farm. For most dairy farms there is no pricing incentive right now to produce 100 percent grass-fed milk, unless it is all direct-marketed under a farm label. This may change in the future as there are several ventures under way to produce grass-fed ice cream, cheese, and other products. There is demand for locally produced raw milk and butter from 100 percent grass-fed cows, although local regulations in many areas prevent consumers from having this option. However, until there is some additional economic incentive to produce 100 percent grass-fed dairy products (zero grain), the number of dairy farms managed that way will be small. A brief discussion of the challenges of zero-grain feeding is included in chapter 4.

# Resources

## General Organic Dairy Farming

Macey, A., ed. 2000. *Organic livestock handbook*. Ottawa: Canadian Organic Growers, Inc. Call (613) 231-9047.

Mendenhall, K., ed. 2009. *The organic dairy handbook: A comprehensive guide for the transition and beyond*. Rochester, NY: NOFA–NY, www.nofany .org.

———, ed. 2009. *Transitioning to organic dairy: Self-assessment workbook*. Rochester, NY: NOFA–NY, www.nofany.org.

Northeast Organic Dairy Producers Alliance (NODPA). www.nodpa.com. This Web site has useful articles on organic management, transitioning to organic, economics, and organic milk markets. NODPA also has an online discussion listserve and quarterly newsletter.

## Organic Standards and Transition

Appropriate Technology Transfer for Rural Areas (ATTRA). This site has a large number of articles on organic and sustainable farming methods. http://attra.ncat.org/organic.html.

eOrganic. This site includes articles, videos, and webinars on organic dairy farming and the organic standards. www.extension.org/organic_production.

Organic Materials Review Institute (OMRI). This organization reviews products to determine which ones are allowed for use on organic farms. www.omri.org.

National Organic Program (NOP). This site has a list of certifiers, the organic standards, and other information. www.amu.usda.gov/nop.

Sustainable Agriculture Research and Education (SARE). www.sare.org/htdocs/pubs. Call (802) 656-0484.

# Livestock Health Care

## General

Barrell, G. K., ed. 1997. *Sustainable control of internal parasites in ruminants.* Lincoln University, New Zealand: Animal Industries Workshop.

Duval, Jean. 1997. The control of internal parasites in cattle and sheep. Publication no. 70. Ecological Agriculture Projects. Call (514) 398-7771.

———. 1997. *Treating mastitis without antibiotics.* Ecological Agriculture Projects. Call (514) 398-7771.

Ellingwood, Finley. 1998. *American materia medica, therapeutics and pharmacognosy.* Sandy, OR: Eclectic Medical Publications.

Holliday, Richard J., DVM. "Fundamentals of holistic animal health." Call (800) 626-5536.

Hoke, Dave, DVM. 1994. *A new troubleshooter's guide to dairy cows.* Hastings, New Zealand: Touchwood Books. Available from NOFA–VT. Call (802) 434-4122.

Jarvis, D. C., MD. 1958. *Folk medicine.* New York: Fawcett. Available from Acres U.S.A. Call (512) 892-4400.

Karreman, Hubert J., DVM. 2004. *Treating dairy cows naturally: Thoughts and strategies.* Paradise, PA: Paradise Publications.

*The Merck Veterinary Manual.* Rahway, NJ: Merck and Co.

Penn Dutch Cow Care. Web site: www.penndutchcowcare.org/.

Sloss, Margaret W., Russell L. Kemp, and Anne M. Zajac. 1994. *Veterinary clinical parasitology.* 6th ed. Ames, IA: Iowa State University Press. Call (800) 862-6657.

Turner, Newman. 1970. *Herdsmanship.* London: Faber and Faber. Available from Acres U.S.A. Call (512) 892-4400.

## Medicinal Herbs

Bairacli Levy, Juliette de. *Complete herbal handbook for farm and stable.* First published 1952. Kent, U.K.: Mackays of Chatham.

Buhner, Stephen H. 1999. *Herbal antibiotics.* Pownal, VT: Storey Books.

Foster, Steven, and James A. Duke. 1990. *Peterson's guide to medicinal plants.* Boston: Houghton Mifflin Co.

Grieve, M. 1931. *A modern herbal.* 2 vols. Reprint, New York: Dover Publications.

Hoffman, David. 1992. *The new holistic herbal.* Rockport, MA: Element Books.

Ody, Penelope. 1993. *The complete medicinal herbal.* New York: D. K. Publishing.

### Homeopathy

Day, Christopher. 1995. *The homeopathic treatment of beef and dairy cattle.* Beaconsfield, Bucks, England: Beaconsfield Publishers Ltd.

Hansford, Philip, and Tony Pinkus. 1992. *The herdsman's introduction to homeopathy.* Reprint 1998. Tunbridge Wells, Kent, England: Helios Homeopathic.

Macleod, George. 1997. *The treatment of cattle by homeopathy.* Essex, England: C. W. Daniel Co. Ltd.

———. 2002. *A veterinary materia medica and clinical repertory.* Essex, England: C. W. Daniel Co. Ltd.

Sheaffer, C. Edgar. 2003. *Homeopathy for the herd.* Austin, TX: Acres U.S.A. Call (512) 892-4400.

### Nutrition

Hedtke, J. L., et al. 2002. Quality of forage stockpiled in Wisconsin." *Journal of Range Management.*

Hoffman-Sullivan, K., R. J. DeClue, and D. L. Emmick. 2000. *Prescribed grazing and feeding management of lactating dairy cows.* Syracuse, NY: New York State Grazing Lands Conservation Initiative/USDA–NRCS.

*Livestock nutrition from field to feeder.* 2003. Homestead Organics. Call (613) 984-0480.

Morrison, F. B. 1959. *Feeds and feeding.* 22nd edition. Ithaca, NY: Morrison Publishing Co.

National Research Council. 2001. *Nutrient requirements of dairy cattle.* Washington, DC: National Academy Press.

### Soil Fertility

Clark, A., ed. 2007. *Managing cover crops profitably.* 3rd ed. Beltsville, MD: Sustainable Agriculture Network, www.sare.org/publications/covercrops/covercrops.pdf.

Gershuny, Grace. 2011. *Compost, vermicompost, and compost tea.* White River Junction, VT: Chelsea Green Publishing.

Gershuny, Grace and Joe Smillie. 1999. *The soul of soil,* 4th ed. White River Junction, VT: Chelsea Green Publishing.

Gilman, Steve. 2011. *Organic soil-fertility and weed management.* White River Junction, VT: Chelsea Green Publishing.

The Josephine Porter Institute for Applied Bio-Dynamics, Inc., http://www.jpibiodynamics.org.

Kinsey, Neal. *Hands-on agronomy.* Available from Acres USA. Call (512) 892-4400.

Kuepper, G. 2003. Manures for organic crop production. ATTRA, www.attra.ncat.org/attra-pub/PDF/manures.pdf.

Lampkin, Nicolas. *Organic farming.* 2002. Ipswich, England: Old Pond Publishing.

Magdoff, Fred, and Harold van Es. 2000. *Building soils for better crops.* 2nd ed. Sustainable Agriculture Network handbook series. Washington, DC: USDA. Call (301) 504-6425.

Sattler, F., and E. von Wistinghausen. 1992. *Bio-dynamic farming practice.* Stourbridge, England: Bio-Dynamic Agricultural Association, http://www.biodynamics.com/.

Soil Foodweb. www.soil-foodweb.com. Dr. Elaine Ingham's Web site on soil organisms.

Sullivan, P.G. 2001. Assessing the pasture soil resource. ATTRA, www.attra.ncat.org/attra-pub/PDF/assess.pdf.

## Crops

Ball, Don, Mike Collins, et al. 2001. *Understanding forage quality.* Park Ridge, IL: American Farm Bureau Federation.

Bowman, G. 2001. *Steel in the field: A farmer's guide to weed management tools.* Beltsville, MD: Sustainable Agriculture Network, www.sare.org/publications/steel/steel.pdf.

Canadian Organic Growers. 2005. *Gaining ground: Making a successful transition to organic farming.* Ottawa: Canadian Organic Growers, www.cog.ca.

Heath, M. E., Robert F. Barnes, and Darrel S. Metcalfe. 1985. *Forages: The science of grassland agriculture.* Ames, IA: Iowa State University Press.

Turner, Newman. *Fertility farming.* Reprinted 1995. Austin, TX: Acres U.S.A. Call (512) 892-4400.

Wallace, J., ed. 2001. *Organic field crop handbook.* 2nd ed. Ottawa: Canadian Organic Growers, Inc., www.cog.ca/ofch.htm.

*A whole-farm approach to managing pests.* Beltsville, MD: Sustainable Agriculture Network, www.sare.org/publications/farmpest.htm.

## Grazing

*Graze* Magazine. For info, call (608) 455-3311.

Murphy, Bill. 2002. *Greener pastures on your side of the fence: Better farming with voisin management-intensive grazing.* Colchester, VT: Arriba Publishing.

Smith, Burt. 1998. *Moving 'em: A guide to low-stress animal handling.* Kamuela, HI: Graziers Hui.

*Stockman Grass Farmer* Magazine. For info, call (800) 748-9808.

Northeast Grazing Guide. www.umaine.edu/grazingguide.
www.behave.net Information on why animals choose to eat certain plants and not others.
Wisconsin Extension. *Identifying pasture grasses* and *Identifying pasture legumes.* http://learningstore.uwex.edu. Call (877) 947-7827.

### Livestock Breeds and Selection

American Livestock Breeds Conservancy. Call (919) 542-5704.
Drayson, James. 2003. *Herd bull fertility: A manual for purebred cattlemen, commercial cattlemen, veterinarians, professors and students of agriculture, anyone involved in the cattle industry.* Austin, TX: Acres U.S.A.
Fry, Gearld, and Charles Walters. 2003. *Reproduction and animal health.* Austin: TX: Acres U.S.A.
New England Heritage Breeds Conservancy. For info, call (413) 443-8356.

## Organic Production and Marketing

### Processing and Marketing Dairy Products

American Cheese Society, Web site: www.cheesesociety.org.
Dunaway, Vicki H. *The small dairy resource book: Information sources for farmstead producers and processors.* Available from SARE. www.SARE.org.
Schmid, Ron, ND. 2009. *The untold story of milk.* Winona Lake, IN: New Trends Publishing.
www.rawmilk.org

### Nutrition and Grass-Fed Animal Products

Robinson, Jo. 2000. *Why grass-fed is best.* Vashon Island, WA: Vashon Island Press. Call (866) 453-8489, or see www.eatwild.com.
———. *Pasture Perfect.* Vashon Island, WA: Vashon Island Press. Call (866) 453-8489, or see www.eatwild.com.
*Wise Traditions* Magazine. Weston A. Price Foundation, www.westonaprice.org. For info, call (202) 333-HEAL.

# Index

# About the Author and Illustrator

Sarah Flack is a sought-after workshop presenter, an agricultural consultant, and independent organic inspector. She was raised on a family farm in northern Vermont, and has worked on organic and nonorganic dairy, vegetable, sheep, and beef farms. She does independent organic certification inspections of farms and processing facilities and consults on grass farming, farm design, farm-business management, and planning. Sarah has a masters in agronomy and has also studied animal science, organic livestock health care, animal welfare, production of value-added meat and dairy products, and farm-business planning. She helped found the Vermont Grass Farmers Association, and currently works with NOFA–VT and other agricultural organizations.

Jocelyn Langer is an artist, music teacher, and organic gardener, and the illustrator of the NOFA organic-farming handbooks. She illustrates and does graphic design work for alternative media and political events as well as organic-farming-related publications. Jocelyn lives in central Massachusetts.

Kathie Arnold and Jeanette Fellows were the special farmer-reviewers for this handbook, and Hue Karreman and Jerry Brunetti were the scientific reviewers.